AUSTRALIA

DOWN UNDER

Dedicated to our children:
Adam, Tammy and Troy
in the hope that they will inherit
a pristine ocean wilderness.

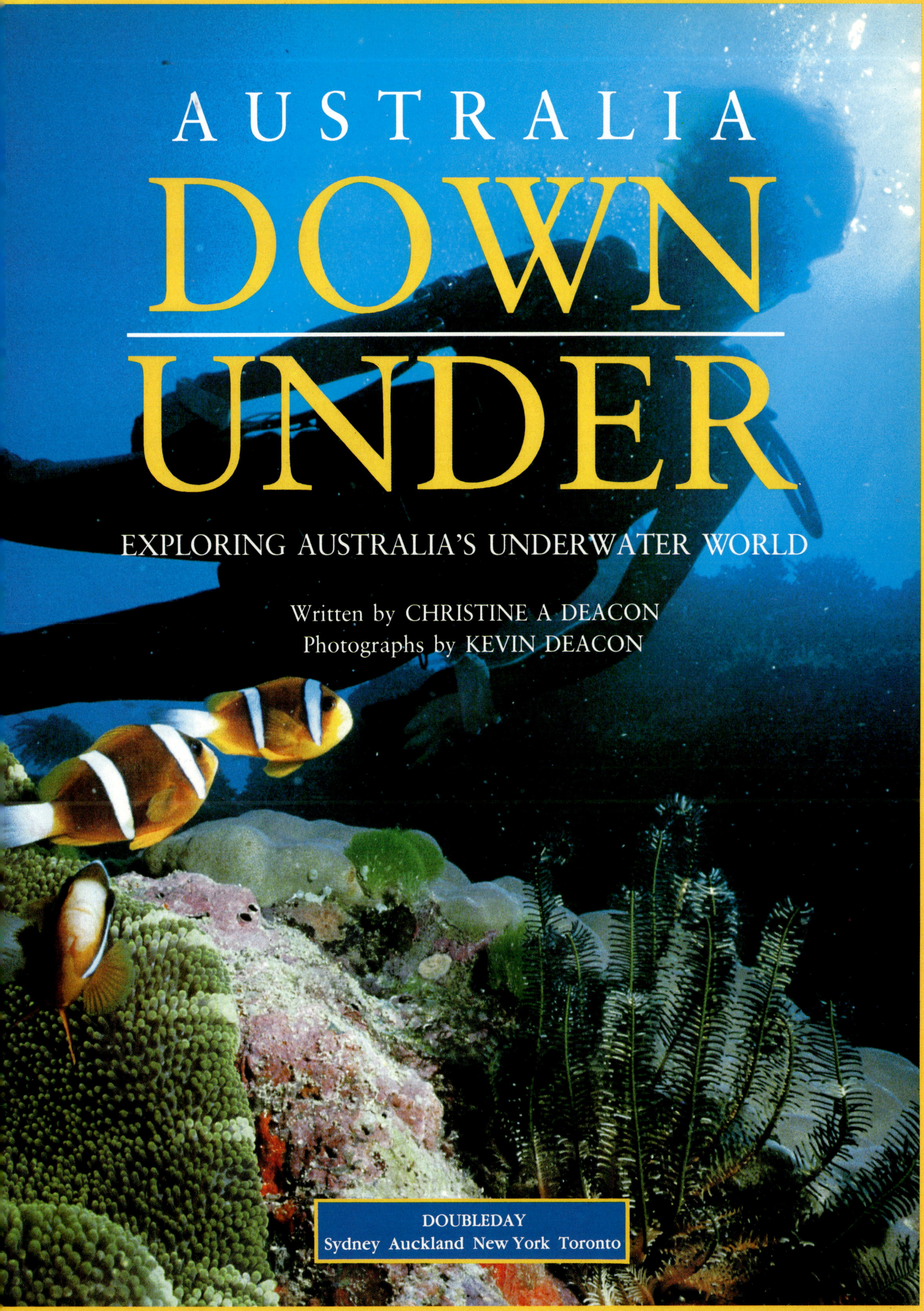

AUSTRALIA DOWN UNDER

EXPLORING AUSTRALIA'S UNDERWATER WORLD

Written by CHRISTINE A DEACON
Photographs by KEVIN DEACON

DOUBLEDAY
Sydney Auckland New York Toronto

ACKNOWLEDGMENTS

A book such as this would not be possible without the effort of many people. I would like to thank them all. I would particularly like to thank the team at Dive 2000, Ian Sizeland, Hugh Scarlett and Kath Savage for taking on extra duties so that Kev and I could spend the considerable time required to select and caption the photographs which you see in this book. In this way Bob Pritchard and Chris Deacon, Kev's second wife, made a very significant and much appreciated contribution, and Chris also appears in many of the photographs. She and Kev provided their experiences without reservation, on the many obscure and difficult questions which I persisted in asking. I would like to thank all those people I interviewed, particularly Ian Lewis of Mount Gambier and Barry May from Cairns. I would like to say thanks to Lois Horton-James for preparing the initial maps and bromide illustrations. Numerous photographers obliged with photographs, which enhance the geographical coverage of Australia — down under. I would like to specially thank Bill and Peter Boyle, Dave Warth and Andy Fraser for their very artistic photographs. Jeff Mullins and Peter Stone of Aquarius Dive Travel provided considerable information on areas I have not dived. Brian Maloney provided photographs of his wonderful paintings and Yak Carrasco allowed us to photograph him in action — painting by the sea.

There is a long step between the conception of the idea for a book and its acceptance by a publisher. I would like to thank Wally Patterson, my agent, Lisa Highton, Kirsty Melville and the designer Susan Kinealy at Doubleday Australia who helped make the dream a reality. I would also like to thank Robyn Williams for his assessment of our efforts.

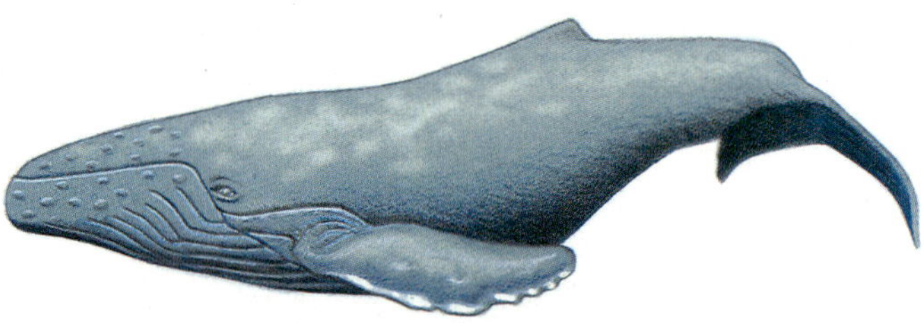

AUSTRALIA DOWN UNDER
First published in Australasia in 1986
by Doubleday Australia Pty Limited Reprinted 1987
This edition reprinted in 1988 by Doubleday, a
division of Transworld Publishers
(Aust.) Pty Ltd, 15-23 Helles Avenue
Moorebank NSW 2170

Designed by Susan Kinealy
Maps by Mike Gorman

National Library of Australia
Cataloguing in Publication data

Deacon, Christine.
Australia down under.

Bibliography.
Includes index.
ISBN 0 86824 241 1.

1. Marine biology — Australia. 2. Diving, Submarine — Australia.
3. Underwater exploration — Australia. I. Deacon, Kevin II. Title.
574.994

Typeset in Sabon by Graphicraft Typesetters Ltd, Hong Kong
Printed in Singapore

FOREWORD

Back in the 1950s I saw the films of Hans and Lotte Hass — some of the very first pictures shown in the cinema about underwater life. Then I thought: 'This must be the perfect existence — cruising on boats in the tropics, diving to explore places of magical beauty, and making worthwhile scientific contributions at the same time. (And you go to work in swimming gear!) I was certain that absolutely everybody would immediately think the same and soon the world would be 100 per cent marine scientists.

But it is not so. Even in Australia it was ages before people took a serious look at what is off one of the longest coastlines in the world. Not until the late 1920s was there a full-scale scientific expedition to the Great Barrier Reef. It was led by Sir Maurice Yonge, the great biologist from Scotland. He brought several first class scientists with him — most of them women. Why the neglect? Was everybody scared of sharks and box jellyfish?

Good scuba gear has not been available all that long. There was much to catch up on in the new southern waters — new to science that is. But the real reason it has taken so long to discover Australia down under is just that: the dominance of northern hemisphere studies. Only now are enthusiasts like Christine Deacon and Kevin Deacon and institutions like the Australian Institute for Marine Science near Townsville helping us catch up. There is certainly a lot of catching up to do as, for example, there are many times more fish and different species in one patch of the Barrier Reef than there are in the North Sea.

Yet even today I hear scientists express mystification at how little Australia does in some quarters. Recently I talked to a professor of zoology in Singapore who had just completed a tour of Australia, meeting our ichthyologists. He was amazed at how few there are in our universities studying fish, especially at a time when we need to obtain basic knowledge about migration, cropping and conservation.

For these reasons, I am delighted to see books like this one trying to introduce marine science to lay people. Straightforward information, told personally, together with superb pictures, will surely inspire people to explore our magnificent marine heritage, both as professional scientists and as amateurs. In many cases it is the amateurs who provide the population studies and the driving enthusiasm. It is to them as well as to the professionals, that we owe the awareness there is of how very much there is to protect.

Robyn Williams

Lord Howe Island is a favourite site for divers.

CONTENTS

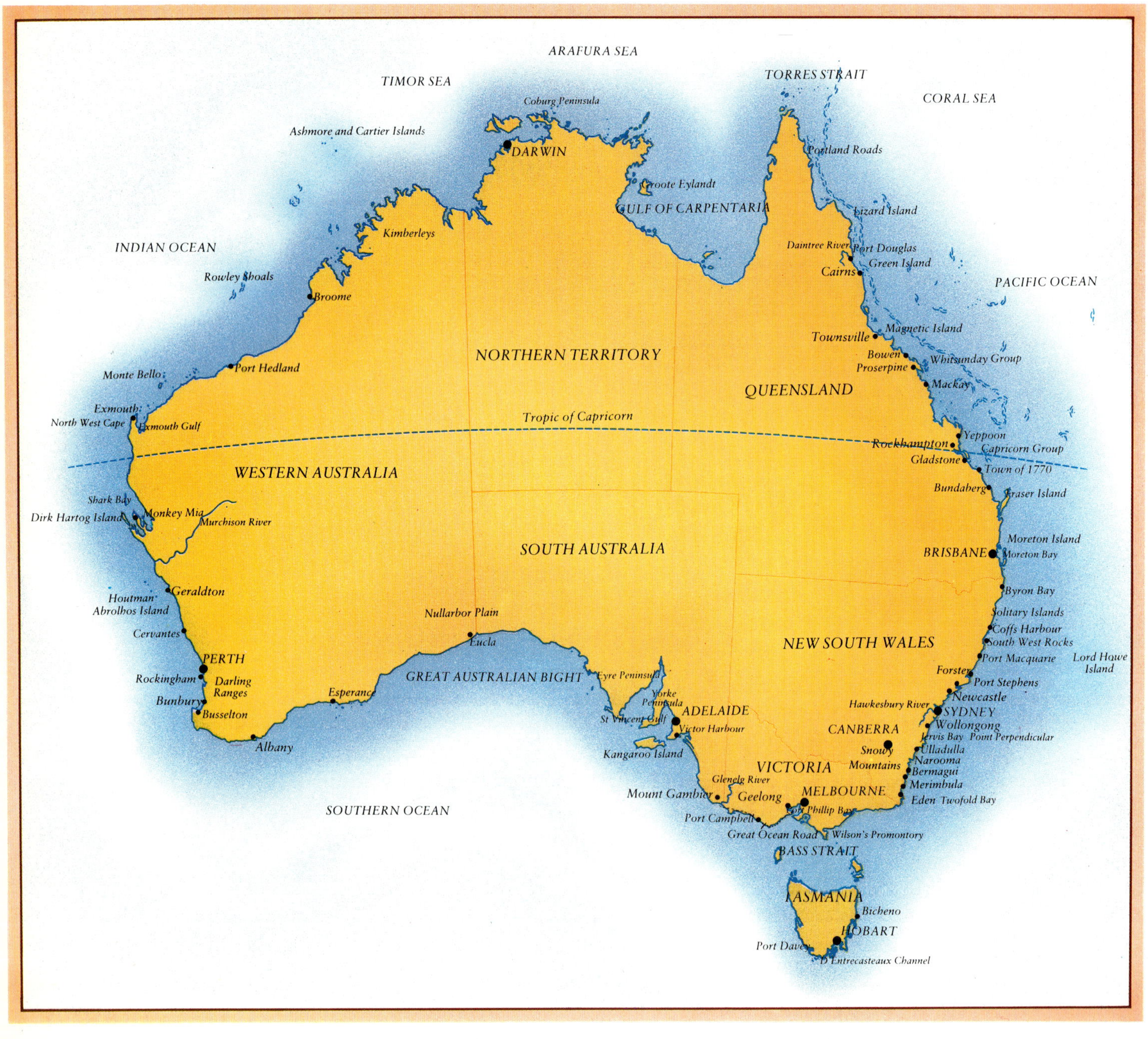
ARAFURA SEA
TIMOR SEA
TORRES STRAIT
CORAL SEA
Coburg Peninsula
Ashmore and Cartier Islands
DARWIN
Portland Roads
Groote Eylandt
GULF OF CARPENTARIA
Lizard Island
INDIAN OCEAN
Kimberleys
Daintree River
Port Douglas
Green Island
Cairns
Rowley Shoals
Broome
PACIFIC OCEAN
Townsville
Magnetic Island
NORTHERN TERRITORY
Bowen
Proserpine
Whitsunday Group
Monte Bello
Port Hedland
Mackay
QUEENSLAND
Exmouth
North West Cape
Exmouth Gulf
Tropic of Capricorn
Yeppoon
Rockhampton
Capricorn Group
Gladstone
Town of 1770
WESTERN AUSTRALIA
Bundaberg
Fraser Island
Shark Bay
Dirk Hartog Island
Monkey Mia
Murchison River
Moreton Island
SOUTH AUSTRALIA
BRISBANE
Moreton Bay
Geraldton
Byron Bay
Houtman Abrolhos Island
Nullarbor Plain
Solitary Islands
Cervantes
Coffs Harbour
South West Rocks
Eucla
NEW SOUTH WALES
PERTH
Port Macquarie
Lord Howe Island
Rockingham
Darling Ranges
GREAT AUSTRALIAN BIGHT
Eyre Peninsula
Forster
Port Stephens
Bunbury
Esperance
Yorke Peninsula
Newcastle
Busselton
ADELAIDE
Hawkesbury River
SYDNEY
St Vincent Gulf
Victor Harbour
CANBERRA
Wollongong
Jervis Bay
Point Perpendicular
Albany
Kangaroo Island
Snowy Mountains
Ulladulla
VICTORIA
Narooma
Bermagui
Glenelg River
MELBOURNE
Merimbula
Mount Gambier
Geelong
Eden
Twofold Bay
SOUTHERN OCEAN
Port Phillip Bay
Port Campbell
Great Ocean Road
Wilson's Promontory
BASS STRAIT
TASMANIA
Bicheno
HOBART
Port Davey
D'Entrecasteaux Channel

INTRODUCTION

The world beneath the sea is the last frontier on earth. In the last twenty-five years we have begun to explore its depths, to learn about its seasons and cycles, and to appreciate and enjoy its inhabitants. In Australia we are lucky to have the greatest marine wonder in the world — the Great Barrier Reef. We have a coastline that extends from the tropics to temperate seas and we have a great variety of underwater scenery and life within that range.

We are extending our knowledge of the seas rapidly. Within my own lifetime the theory of continental drift has been proposed to explain how the continents and oceans were formed. When I first heard of the theory, I thought, how could it be that huge continental plates were moving like leaves floating on the earth's fluid core? The idea was so foreign, I dismissed it, but now the majority of the scientific community accepts the theory as the explanation for the formation of the continents.

The theory of Continental Drift taught us about the two supercontinents, Gondwanaland and Laurasia. The original continents that existed 1000 million years ago, they were situated in the southern and northern hemispheres respectively. In this context we will be focussing on Gondwanaland. The similarity of the rocks of South America and southern Africa was demonstrated by the geologist Alex L. duToit in 1937. He fitted the coastlines of the continents together, noting equivalent formations on opposite coasts with regard to their origin and subsequent history. But it was not until the evidence from magnetic studies of the ocean floor and ridges was established that the theory began to be accepted.

In the study of the earth's magnetism, it was found that the magnetic field of the earth had reversed at specified periods over the last 3.6 million years. The patterns of magnetism in the rocks on the ocean floor showed a striped pattern. The discovery that the pattern was symmetrical on either side of an ocean ridge was made by F. J. Vine and Tuzo Wilson. What happened was that each time the molten material, which gradually rose from the mid ocean ridges, solidified, it adopted the magnetism of the earth. As the earth's magnetism reversed, so the rocks recorded the changed magnetism.

Once the mechanism for the creation and spreading of the seafloors had been established, the geological evidence had to be examined to see if the continents fitted together. Matching the geological provinces in South America and Africa showed remarkable continuity. Evidence from Antarctica, India and Australia confirmed the fit, when the depth of the surrounding continental shelf was taken as the shoreline. Further evidence came from the presence of the plant *Glossopteris*, which occurred in coal deposits in Antarctica and other Gondwanaland continents.

Ocean circulation patterns have been shown to have far reaching effects on earth. Studies of the ocean enable us to have a holistic approach to studies of the earth.

The oceans influence climate and are influenced by climate in many ways. Water temperatures offshore influence the rainfall. The temperature of bodies of water affect their flow and mixing at the surface and down through the water column The rotation of the earth accounts for the rotation of the major ocean gyres or circles, which serve to transfer heat from the equator anticlockwise down the western ocean – land boundary, as the east Australian current does. The cooler water flowing through Bass Strait, is part of the West-Wind Drift. It joins the gyre to flow across the Pacific and turn north with cooler water off the coast of South America. In the Indian Ocean the West Australian current is a warmer south flowing current influenced by the monsoon which also leads to fluctuations of currents through Torres Strait to the north of Australia. Winds are another factor in the flow of ocean currents. Strong winds from the west at mid-latitudes and east winds at low latitudes near the equator, promote the turning of the gyres. Wind driven currents are called drifts and in contrast to the winds are named after the direction of their destination, thus an easterly current is blown by a westerly wind.

So far we have mentioned only currents on the surface, but there are currents mixing the ocean vertically. Surface water sinks as it becomes more dense from cooling or increases in salt concentration caused by evaporation. Where surface flows converge as at the centre of a gyre the water also subsides as at the Antarctic Convergence to the south of Australia. The seas on each side of the convergence differ by about four degrees centigrade and create a barrier preventing mixing of seas of different temperature and salinity.

A picture of the ocean circulation around Australia has begun to emerge with the introduction of satellites to the study of ocean circulation. Large eddies have been discovered rotating anticlockwise off the east coast of Australia. These eddies have a diameter of about 200 or 300 kilometres and a surface speed of three to four knots. Each eddy may have a life of a year or more as it drifts southwards, but sometimes it may stop for months or even move northward. The eddy may retain its structure to 1500 or 2000 m deep but the currents associated with the eddy are strongest at the surface and may be two degrees warmer than the surrounding water in late winter. An enormous amount of energy is stored in an eddy, about 1 per cent of Australia's annual energy needs, but we have not learned how to tap it. The strong circulation in the deep ocean, affects the continental shelf — the extension of the land's fringe, beneath the sea. Off the New South Wales coast, the main southerly flowing East Australian Current with its eddies veers away from the coast towards the south-east forming a vast front which is more pronounced in summer than in winter. Zonal jets north and east of the separation front may travel as far as New Zealand. Where these eddies and the surrounding sea abut, there is significant nutrient enrichment and this has led to phytoplankton and zooplankton abundance. Further studies on the biological components of these eddies showed variety between species and in size of particular species. The surface temperature fronts which mark the boundaries of the eddies are often highly productive and this has been used to advantage by fishermen at Eden. Regular aerial surveys of these fronts were used to guide the southern bluefin tuna boats to fish, but until the discovery of the eddies, and the upwelling associated with them it was not realised why this was effective.

In Western Australia, a north flowing West Australian Current had been observed by ships' navigators, but the upwelling usually associated with ocean gyres on their eastern edges was absent in Western Australia. The eddies off Western Australia in contrast to those on the eastern Australian seaboard, were both cyclonic and anticyclonic (or anticlockwise) and are embedded in a southerly current for much of the year. Two water bodies mixing off Fremantle generate these eddies. During late autumn, quite suddenly, warm fresh tropical water moves down the coast of Western Australia along the continental slope and the continental shelf, following the continental slope around Cape Leeuwin into the Great Australian Bight. It is known as the Leeuwin Current and its annual appearance in late autumn is thought to be associated with the monsoons in Asia. This current has important consequences for the western rock lobster fishery, the major export fishery in Australia, and for the tuna fishery. Tuna are swept south by the warm Leeuwin Current from their breeding areas in the north. The current boundary where the tuna congregate conforms with the edge of the continental shelf and is an area of upwelling. Eddies spinning off from the Leeuwin Current trap tuna in them and are valuable fishing sites. Identifying the eddies by using satellite maps which show sea surface temperature is a practical means of increasing catches and these maps have been made available to fishermen in Albany and Esperance.

The continental shelf is the relatively shallow platform around the continents. At its seaward edge the seabed suddenly slopes more steeply into the deeper ocean basin. The continental shelf is important economically, for its resources are readily available. Fisheries as well as minerals and oil are being exploited around the 2.6 million square kilometres which comprise the continental shelf around Australia.

Australia was one of the first nations to sign the new Law of the Sea Convention in 1982. Australia had economic and strategic interest in the negotiations and successive governments have given high priority to the negotiations over an eight year period. The territorial sea covers 12 nautical miles and the 200 nautical mile Exclusive Economic Zone gave Australia an additional million square nautical miles of territory, rich in mineral, oil, and marine resources. It also gave freedom of communication, and commercial and cultural links which lie over the sea. The technological innovations this century spurred on an increase in demand for the sea's

Sand and shingle reflect the ripples of surface waves as a school of fish are bathed in sunlight.

Divers on a remote coral island setting out for the day's diving.

resources and led many nations to call for a conference for a new Law of the Sea. Several governments including the Australian government, sought a conference in the General Assembly of the United Nations, to determine the maximum breadth of the territorial sea and the related questions of passage through straits, resources and pollution. It took years of painful negotiation to determine the balance of rights and obligations.

Australia has established the Australian fishing zone as the first step in exercising its claims to the 200 mile Exclusive Economic Zone. This came into operation in 1979. Other nations had already declared jurisdiction including Japan, the United States, the Soviet Union and New Zealand. Australia was cautious in extending its jurisdiction because of the need to reach suitable arrangements with the major fishing interests in the region — those of Japan and Taiwan — to avoid the major surveillance and enforcement problems which the size of the zone created.

Australia opposed the limitations on the continental shelf, to a maximum of 350 nautical miles, claiming its rights over the continental shelf extended out to the edge of the continental margin (consisting of the shelf, continental slope and rise). Australia has a number of ridges and rises forming part of its continental shelf, including the Lord Howe rise and the Norfolk ridge. It is considered the formula derived from the discussion that will not have any impact on these areas. Although Australia has signed the new Convention there has been no indication whether it intends to adopt the continental shelf definition that the outer limit of the continental shelf shall not exceed 350 miles.

Australia was concerned over one other issue relating to the continental shelf, that of historic shipwrecks. In the 1976 Historic Shipwrecks Act, Australia asserted control over wrecks on its continental shelf.

While the negotiations proceeded over eight years, final agreement was reached on the delimitation of the EEZ/continental shelf between nations with opposite or adjacent coasts, by agreement on the basis of international law. Australia is expected to pursue a cautious approach to its maritime claims, reflecting the provision of the Law of the Sea Convention, which it worked long and hard to achieve.

The environment which the Law of the Sea negotiations aimed to protect includes the mid water as well as the shelf environment. Many animals and plants form the ecological associations of the different geographic areas. The tropics north of the Tropic of Capricorn have scattered coral reefs, some extending south of the Tropic. Corals, the microscopic animal which makes the skeleton we know as coral, have built the structure of these enormous reefs. Coral reefs have existed in the shallow seas of the earth for at least 450 million years, though they have changed considerably from the ancient rugose corals. The scleractinian coral which are related to the corals we know today, probably evolved in the Tethys Sea which divided the original supercontinents of Laurasia and Gondwanaland. This sea eventually closed as Australia was moving northwards with continental drift, and it came into the region of warmer water where coral growth was favourable. Coral requires warm water temperatures and abundant light (for the algae which live in the tissues of the coral), so coral growth is favoured by warm shallow tropical waters. The Great Barrier Reef is relatively young. Since its development there have been great variations in sea level, at times the reef has been above water and cut by river channels. Evidence of these old sea levels is seen by scientists when they examine the sediments with seismic refraction. They have found old river beds in the channels or passages through the reef. Most of the present reefs were built on older reef bases when the continental shelf was again flooded as the sea level rose. At the limit of the range of coral growth, in the Abrolhos Islands in the west and Lord Howe Island in the east, the coral growth is limited by temperature, with light availability a secondary limitation. Excessive algae growth in summer may limit the light available. The Abrolhos Islands are relic karst limestone with blue-holes and small lagoons. These karst landscapes are found again on the Nullarbor, where they have been weathered to form the caves which divers explore. The limitation of coral growth has prevented full development of these Abrolhos Islands reefs which would have obscured these earlier karst features. Corals occur encrusting the rocky reefs as far south as Perth and Sydney but they do not attain reef development. The southern waters of Australia are characterised by the presence of algae growing attached to the rocky reefs where the water is shallow and light levels are sufficient to allow photosynthesis. Below this level sponges and other encrusting animals like sea squirts and gorgonia corals which are not limited by light are found attached to the rocky reefs. While the number of species is more limited in southern waters, huge numbers of a single species may be present. The planktonic marine life floating on the surface and other organisms in the water column reach their greatest development in the south and the huge plankton-feeders, the whales and whale sharks are found congregating near this food source.

The Aboriginal people, the first Australians, who arrived 40 000 years ago had an economy based on the resources of the sea. Europeans explored these shores from the 17th century. The earliest landing which is fully authenticated is that of the Dutch ship *Duyfken* in 1606 when William Jansz explored Cape York Peninsula. There were almost certainly visitors to our shores before this. A period of intense exploration of the west coast of Australia by the Dutch ships followed and many of them were wrecked on our shores. They travelled along the trade wind route and the difficulty they had in measuring longitude accounted for their arrival on the coast before they anticipated it. They were wrecked often without warning, sometimes in the middle of the night. Some survived to recount their adventures, others inevitably did not, so there are probably still ships from this period which will be found as we explore the coast and continental shelf with more advanced techniques and more sophisticated equipment.

We are beginning to appreciate the important part which the sea has in our lives, in our history and in our future. The appreciation of the marine world is more difficult than appreciation of unspoilt wilderness on land. Very few people have seen beneath the surface and those who have gone down under, have seen only a small part of it, for we are simply scratching the surface. The inter-relationships between the many parts require a systems approach to the management of unique features. It is important that people are aware of the beauty and potential of the marine world. Education is I believe the way to increase awareness. It is for this reason that our marine reserves should be created with multi-purpose planning, so that we conserve their unique life and inhabitants and awaken Australians to their importance.

This book begins with an introduction to diving, which is one of the best ways to appreciate the world down under. It explores the regions and the different diving activities. Each chapter starts with an experience which shows an aspect of diving which relates to that chapter. The dangers of diving are minimised by correct procedures and equipment, but it has not always been so and experiences of some encounters with dangerous marine animals have been included. The appendices list dive shops and dive sites that are mentioned in the text. Finally there is a list of safety precautions and first aid techniques.

Your plunge into another world, whether you do so from the comfort of your chair, or by donning mask, fins and snorkle will increase your awareness of the sea. We hope that everyone who reads this book enjoys the experience and has a greater awareness of Australian seas and the need to conserve our underwater heritage.

Charter boats anchor on coral reefs by dropping anchor on top of the coral bommie where the coral will not be damaged.

EXPLORING UNDERWATER

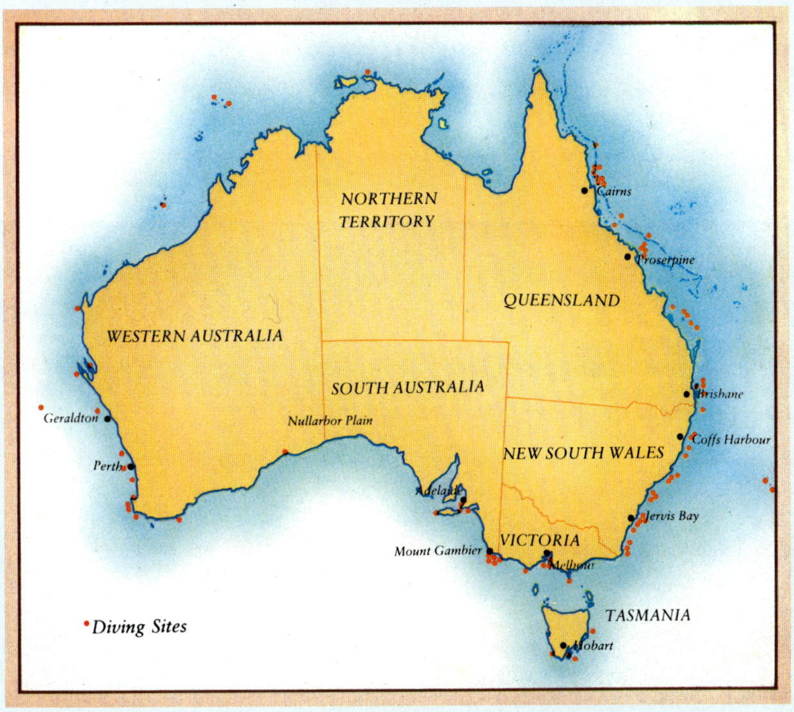

Whether you wish to explore Australia's underwater world from your armchair or by diving yourself, it is helpful to have a background to the nature of Australia's marine life and where it lives. From the warm tropical waters at the tip of Cape York to the cold forbidding waters off Tasmania, there is a wide spectrum of conditions and habitats in which marine plants and creatures survive. There are open sand plains, extensive seagrass meadows, dense seaweed and kelp jungles, as well as tropical and temperate reefs.

Australia's seas contain one of the earth's greatest concentrations of marine life. Almost 30 per cent of all marine fish in the world (approximately 3500 species) as well as many other curious and colourful creatures inhabit the waters off the Australian coast. As each geographical region changes from north to south, so too does the life within each ecosystem. The boundaries between the regions are determined mainly by variations in currents, water temperature and geological formations.

Of all these underwater environments, coral reefs are the richest. The Great Barrier Reef, which is made up of 2500 individual reefs, is the largest reef system in the world. Yet it is the reefs and seagrass meadows of the southern Australian region that are among the world's leaders in diversity of underwater life. The great majority of the fish that inhabit this temperate region, though less colourful than those that inhabit the continent's tropical waters, are found nowhere else in the world.

It is this uniqueness, together with the enormous variety of underwater life around the coast and inland that make exploring down under so fascinating.

"My first experience of diving was a long time ago. It impressed me as a novelty impresses a child. I was fascinated as I watched diver Harold Chadwick climb into the rubberised canvas suit, his attendant assisting him to pull it over his woollens, then fasten the brass boots which I could not lift. Breastplates of lead were tied to his back and chest and the copper helmet with the gridded brass windows was screwed down. The pump began its noisy drone and Harold descended the ladder of the barge beneath my father's yacht Thara *to check the mooring. He found another mooring that had sunk, and duly surfaced for a wire cable to fasten it to the winch. Slowly the booty of the sea came up through the clear water, covered in weed and shells. Harold surfaced and regaled me with tales of what he had seen below. I was hooked and from that day to now I have been fascinated by what lies down under."*

Traditional hardhat dress is still used occasionally on commercial diving sites. The US navy model copper helmet is one of seven models of standard dress which were in use. Today modern, light weight, high impact, plastic helmets and light weight clothing have largely replaced the older style dress.

In this chapter we discuss how to go diving, the equipment used and the many opportunities diving offers. Most people learn to dive for pleasure because diving is fun, exciting and adventurous. Diving is also a rapidly growing industry. For some divers, their sport may become their profession, so the chapter concludes with a brief look at different commercial aspects of diving.

Whatever you decide to do in diving, nothing compares with that first plunge beneath the surface. It's a whole new world from the very first moment: you hear your bubbles gurgling to the surface, and you are suddenly aware of all the sights and sounds of the sea, which you never knew existed as you swam or sailed above. The sensation of weightlessness is probably comparable to flying or floating in outer space, as you glide to the bottom and explore the rocks and ledges of the underwater world.

I finally learned to dive when I turned eighteen. Diving was a new sport and there were few who ventured below and very few were women. Last year 20 000 Australians took up the sport of scuba diving and thirty per cent of them were women. Diver training has attained professional status and recognised certification by one of three professional training bodies is essential in order to buy equipment or to have scuba gear filled with air.

The Professional Association of Diving Instructors, PADI, is recognised in eighty-five countries, has 530 instructors and has trained 200 000 divers in Australia. The Federation of Australian Underwater Instructors, FAUI, is based in Western Australia and has 400 instructors around Australia. It is affiliated with the World Underwater Federation based in France and recognised in fifty-four countries. The National Association of Underwater Instructors, NAUI, originated in the United States of America and has been established in Australia, with its base in Queensland, for three years.

As the diver descends a fish comes out to defend his territory against the intruder.

Instructors offer a variety of training experience including theory, pool training and ocean diving. Teaching aids include audio-visual presentation, training manuals and the latest diving equipment. You can start with a guided tour of a coral reef, whilst on holiday, or you can begin at home, so that you are better prepared for adventure on your holiday.

Training teaches confidence as well as technique. Courses progress by logical steps: from shallow water to deeper water, from pool to ocean. Divers learn to glide over the bottom in a controlled and relaxed way, to handle any situations that might arise, such as a leaky facemask, and to enjoy themselves.

Instructors in basic diving usually take their pupils to where the water is calm and the danger from marine life or pounding surf is minimal. I was once with a group of dive school pupils in a calm, sandy area looking for something to show them on a barren bottom, when I saw an outline in the sand. Thinking it was a flounder, I moved the sand around it to expose it to the students. A bolt of electricity suddenly moved me half a metre, off the bottom. The numb ray stayed in the same place visible when the sand partially moved as it gave me a charge. The charge was stronger than a bolt from an electric appliance that travels up your arm before you feel it on your body, it is more like being hit on the head. The pupils spent the rest of the dive two metres off the bottom. It is not only the dangers that pupils learn to appreciate on their basic diving course, they are also taught how to choose equipment and what to do in emergencies.

Equipment is provided during training. When you are ready to buy your own, a full diving outfit costs about $1000. The equivalent would be $25 a day to hire. Divers often buy their personal diving equipment before they learn to dive and buy their scuba gear later.

Above: PADI and Dive 2000 instructor Hugh Scarlett demonstrates the techniques of brakes and releases to an attentive group of students learning rescue diver techniques in Warringah Aquatic Centre. Kevin Deacon is recording performance values and results as the class progresses.

Below: Dive Master Bob Pritchard practises equipment exchange with a trainee dive master in the pool.

The first item a diver buys is a mask: it must fit well for comfort. To test the fit, place the mask against the face without the strap in place. Inhale through the nose and the mask should stay in place. You should be able to squeeze your nose to clear your ears as you ascend. The snorkel attaches to the mask strap. It should have a comfortable mouthpiece and a large diameter. Fins come in full-foot or open-heel, in synthetic silicon or rubber.

The wetsuit is an important personal item of equipment. Heat loss when breathing compressed air is considerable and wetsuits are essential in cooler climates and in deeper waters, even in the tropics. As well as providing warmth, they protect against abrasions: coral cuts can be painful. In the tropics, divers often prefer Lycra suits which offer protection without being restrictive. Another reason for wearing a wetsuit is to prevent sunburn: sunburn on the back of the legs can be very uncomfortable. Wetsuits should fit snugly for maximum warmth. The retention of body heat by a wetsuit depends on a thin layer of water between the wearer and the suit. The thinner the layer of water, the quicker the body warms it. Wetsuits are buoyant and to counteract buoyancy, lead weights are worn. The weights are worn on a belt that unbuckles quickly in an emergency.

Scuba equipment consists of either a steel or aluminium tank that holds air under pressure. Since tanks are high-pressure vessels, they are subject to strict standards of manufacture and require regular testing and inspection. In Australia, steel tanks must be tested every year and aluminium tanks inspected alternate years and tested every second year. Aluminium tanks are guaranteed for the lifetime of the original owner. The working pressure, the quantity it contains, is less with a steel tank. A tank is fitted with a boot to stand it upright if the bottom is rounded, and with either a J-valve or a K-valve. A J-valve has a reserve activated by raising a lever before diving and releasing it when breathing becomes difficult, the danger being that the lever may be accidentally bumped and the diver could run out of air. It is safer to use a K-valve with a submersible pressure gauge that tells you how much air you have in your tank at a glance.

Divers always explore in pairs, so in the unlikely event of one running out of air, it is possible to share your companion's air. You learn to do this in your basic scuba training.

Divers at a Sydney beach bracing each other and entering the water backwards and sidewards — the correct technique for a beach entry through the surf.

Instructors and dive leaders usually have an additional regulator attached to their tank so that they can share with pupils if they run out of air. This is called an 'octopus system'.

Your tank will also have a backpack that fits comfortably on your back with quick release straps. Single-hose regulators supply air to the diver, at the required pressure, in two stages: the first stage reduces the high tank pressure in the scuba tank to slightly above that of the surrounding water; and the second stage, in your mouth, supplies air as you inhale. A purge button on the front of the second stage enables you to expel any water in the regulator. The regulator may have attachments for other devices such as an octopus regulator, submersible pressure gauge and a buoyancy control device.

Buoyancy control devices or BCDs look similar to lifejackets but they are much more than that: they allow you to glide effortlessly over the underwater terrain when used correctly. They can be inflated manually or from an attachment on the regulator. They are deflated again when ascending by raising the manual inflation mouthpiece above the head. There are several types of BCDs. The earlier ones were worn on the diver's chest but the most popular types now are the back-mounted and jacket styles. A diving watch, depth gauge and compass are other items that make diving safer and more enjoyable.

Divers are snorkelling around dive charter vessel Sea Trek. Her twin keels and shallow draft are ideal in coral reef waters. Charter yachting and diving around the Whitsunday Group, Queensland is very popular. You can sail and dive or request a skipper to help you explore the area. Divers usually snorkel the reefs as space for compressors and tanks is limited on charter yachts.

Diver Propulsion Vehicles or DPVs open a new age of diving. Where divers once had to swim to explore the reef, consuming a lot of air and covering a short distance, now they can drive a DPV. As the time a diver spends underwater is limited to air carried in his scuba, the arrival of DPVs has extended the diver's range. In one dive using a DPV, the diver, instead of exploring a few hundred metres of reef, can now explore kilometres and return safely to the dive boat. A DPV can carry the diver up to two kilometres. Since the diver must get back to the dive boat it is wise not to extend the area from the vicinity of the diving operations. A DPV user can cover a great area at a speed of two knots or he can go slower if he wants to observe the scenery at a suitable pace. As the diver is not expending any energy, he is using less air and the time he can spend underwater, as well as the distance travelled, is increased. Like a short torpedo with handles on either side, the DPVs are compact and simple to use. They provide a sensation of flying as you zoom along the coral reefs, gliding over canyons and drop offs — an experience not easily forgotten. On a wreck, a DPV allows the diver to see the overall picture. While DPVs add interest and safety to dives they also require some specialised training to ensure dive planning and procedures are adequate to prevent problems that can occur with the increase in the diver's range and bottom time.

We dived with DPVs at The Arch, which is directly out to sea from the lighthouse at Point Perpendicular, Jervis Bay. When you anchor on the seaward side, you can dive in water from fifteen metres to thirty metres deep. The Arch is on the deeper stretch parallel to the cliff and is shaped like an aqueduct or Roman bridge. It is in thirty-five metres of water so exploration time is limited, as the diver is using more air at that depth.

We hung our DPVs on a line over the side of the boat and entered the water. When we started our DPVs, we drove down to the bottom in a slow spiral to avoid problems of equalising our ears. When we reached the bottom in thirty metres, we zoomed over to The Arch and within seconds had completed a circuit over the top of it. We signalled to each other to drop down and pass through. We switched on our lights triggered on the handle on the left, and dropped into the deeper water, propelled through The Arch by our DPVs. Our headlights illuminated the eyes of thousands of little fish. We shone our lights on the underside of The Arch and the anemones there shone like jewels glowing in many colours. After a few moments we explored the area beyond The Arch and within five minutes we had seen the area usually covered on a dive here. We were left with time to explore the area around The Arch, which we had never previously had, having always used up our bottom time on The Arch itself. We found The Arch was part of a deep water ledge that extended beyond this area for some distance with beautiful corals, sea whips and fish illuminated in the lights of the DPVs. When our bottom time was half over, we turned and sped back to the anchor with time to spare. When you swim through The Arch from the seaward side you come against a sheer face which opens more to the right near the cliffs. On the ocean side the bottom is sand then reef. We had used only a fraction of the air we would normally use, and had explored a large area in comfort. Deeper dives and more advanced exploration become safer and more efficient with the DPV and the enjoyment of the dive is increased. We surfaced, again in a slow spiral so as not to come up too fast. On the way up a school of metre-long kingfish came in and circled around us. We were able to speed around with them and although we couldn't reach their speed, we were able to experience the sensation of swimming with huge school fish. Hanging our DPVs on the line below the boat, we did a safety decompression stop at three metres. The dive showed us the possibilities of exploring the forty kilometres of cliff and reefs of the Jervis Bay region.

Submarines allow you to explore even greater expanses whilst protected from predators. Scientist Dr Walter Starck, who invented the mixed gas electrolung equipment, reports that sharks have attacked his submarine a number of times, and it is one particular reef shark that is usually involved: the grey reef shark about two metres in length. It is territorial and when the living space around the fish is broached by the submarine, it attacks and often the occupants are unaware of the shark until it strikes the submarine. When exploring a large area of reef there is a chance a shark will attack when the submarine comes upon it unexpectedly.

Before you begin your diving course you will need a medical certificate from your doctor. Some doctors specialise in diving medicine such as at the Diving Medical Centre in Sydney's Neutral Bay. The Centre was established in 1972 to fill an increasing need by amateur divers for specialist medical skills. Since the Diving Medical Centre opened, more than 10 000 divers have had examinations which are recorded at the Centre. It also conducts research into diving medicine.

A professional dive shop is the best place to find out about diving courses in your local area. They often have active dive clubs that offer you the opportunity of diving the best localities as well as obtaining advice on equipment. There are full-time and part-time diving courses available from most professional dive shops and if you work unusual hours, they may even be able to organise an instructor to cater for your special needs. Professional dive shops provide the back-up support that new divers need. They give advice on the latest diving equipment available and are only a phone call away when you want to book a day's diving. There are over 200 professional dive shops in Australia, at least one in every major city or town.

Our coastal waters are accessible to shore diving. It is possible to swim around the rocky headlands of our beaches or explore the quieter waters of our lagoons and estuaries. Divers have been doing so for about twenty-five years and have discovered some very interesting territory in that time. Although many of the best spots are hard to reach, which is why boat diving is so popular, you do not even need to own a boat. Diving trips to many of the best spots are organised every weekend. Your professional dive shop will undoubtedly be able to organise boat diving in your area. Fast runabouts take divers to localities near the major cities and you can dive one site in the morning and another in the afternoon. Charter boats go further afield to the offshore reefs, while personal transportation is provided by inflatable dinghies with outboard motors. Divers made them popular and now the surf clubs use them for search and rescue on our surf beaches.

Instructor Kathie Savage is using a tekna diver propulsion vehicle or DPV to explore this tunnel shaped cave in 15 metres of water.

Boat diving in Australia varies from State to State. In north Queensland, in localities a short distance from the Great Barrier Reef, both day trips and longer weekly or monthly expeditions

Dr Walter Stark's Minisub which is launched from his research vessel, El Torido. *Kevin Deacon is about to climb into the cockpit after shooting movie film on the reef. Minisubs are a considerable advantage when working underwater as the diver's air lasts longer and he can cover long distances while protected from large predators. However, underwater scooters are a less expensive alternative for extending a diver's range.*

Above: Sub See submersibles allow tourists to see the underwater world which divers experience.

Below: Watersport, *a charter vessel that operates from Townsville and goes to the* Yongala *wreck site.*

leave from the larger towns. In Cairns both types of operation, the larger charter vessels and the fast high-speed runabouts, cater for the large number of tourists visiting the area, particularly in winter. Overseas visitors comprise more than fifty per cent of the divers visiting Cairns. Regular flights from the United States of America and charter flights from Japan bring in many divers to operators such as Barry May, who operates several charter vessels from Cairns. These operators have built boats especially for the diving market. Barry has also designed a platform and submarine-type vessel called Sub Sea, which enables non-diving visitors to experience the Great Barrier Reef. There are now five Sub Seas operating on the Queensland coast. From Cairns and Port Douglas, visitors board a fast catamaran and, in less than one hour, are at Low Isles or Hastings Reef where they board the Sub See from a pontoon anchored beside a coral bommie or underwater clump of coral. The pontoon has an underwater viewing chamber where divers can be seen snorkelling over the coral until the Sub See moors alongside. Then it's down the companionway, fold down your seat and sit back for the ride of a lifetime, as you cruise along canyons and mountains of coral, viewing through Sub See's glass windows, the corals and fish of Australia's Great Barrier Reef.

There are other Sub Sees operating at Hardy Lagoon in the Whitsunday islands, reached from Shute Harbour, at Heron Island in the Capricorn group of the Great Barrier Reef Marine Park, and another operating from Townsville. These vessels are very popular and soon many of the resorts and operators along the Queensland coast will have Sub Sees.

The Whitsunday group of islands and resorts are accessible from Shute Harbour near Proserpine. Air Whitsunday can fly you to the outer Great Barrier Reef from Airlie Beach near Shute Harbour in a Buccaneer three passenger seaplane. Divers have the option of travelling out to the reef by seaplane from Shute Harbour or Hayman Island where you can stay in comfort.

Another option available to sailing divers is to hire a yacht. Bareboat chartering means the use of a boat without crew, and you can sail wherever you wish within the boundaries of the charter. Yachtsmen have been enjoying the waters of the Whitsunday group for decades. It's a lot easier to charter a yacht from Mandalay Point, near Airlie Beach, or from Shute Harbour than to sail up the coast from Sydney or Melbourne. The yachts have accommodation for up to eight people and a small compressor and tanks can be included, or you can snorkel among the reefs and sheltered coves as part of your Whitsunday experience.

Further south are the reefs of the Capricorn Group, out from the coastal towns of Gladstone and Bundaberg. Fraser Island, the largest sand island in the world is reached from the coastal town of Hervey Bay.

Diving south of Great Barrier Reef waters is dominated by day boat diving. From Brisbane, day or weekend trips to Moreton Island or day trips to Julian Rocks at Byron Bay, are popular excursions. Coastal New South Wales towns offer weekend and some weekday departures to such diving areas as the Solitary Islands, from Coffs Harbour, or popular sites at Forster, Port Macquarie or Port Stephens. The local professional dive shop will put visitors in touch with local boat operators. As diving grows, more boats are operating out of the smaller coastal towns.

In Sydney, there are twenty professional dive shops operating the fast, stable boats, about seven metres in length, that cater to small groups and are operated under the jurisdiction of the Maritime Services Board. One of the most popular diving areas within easy reach of Sydney is Jervis Bay. Several operators run day trips to the popular sites in this area.

South of Sydney, boats operate dive trips to popular sites from Wollongong, Ulladulla, Narooma and many other towns. At Eden, and over the border at Mallacoota in Victoria, commercial divers harvest abalone, a shellfish popular in South East Asia. They have explored and located many interesting diving areas that sports divers now visit.

Port Phillip Bay is also popular with divers from Melbourne and day boats operate to popular sites near Port Phillip Heads and to Pope's Eye in the middle of the bay. Strong currents restrict the diving to slack tide in most places, which considerably alters the diving conditions and requirements in this area. The southern coastline from Melbourne to Adelaide favours small, fast day boat operations and Victor Harbor is a popular venue from Adelaide.

West Australian divers favour both charter boat and day trip operations. Charter boats operate out to the Abrolhos Islands from Geraldton and boats operate to Rowley Shoals while Perth divers can easily reach Rottnest Island by day boat.

Typical of the catamarans operating out of Sydney is the boat Dive 2000, *seen here being launched.*

While most boat operators run a dive operation catering to qualified divers, some combine training programs which add interest to their operations. Once you have explored your local area and are ready to try something different, advanced diver training will prepare for your expanding interests in the world down under. Advanced diver courses cover navigation, deep diving procedures, techniques for search and recovery, all of which improve your diving skills around practical diving application. A rescue diver course equips you to help someone in distress by training in first aid, resuscitation and problem recognition and solutions. You can also do an equipment course. Divers often have minor problems with their gear which are easily fixed if you know how.

Night diving requires additional skills such as: navigation, equipment handling, and recognition of light signals which you can learn all about in a night diving course while also having a lot of fun. The course introduces another aspect of the sea and the creatures you'll see only at night. These animals are often colourful and spectacular. Initially, divers might be afraid of the dark, unknown waters but night diving provides an adrenalin boost as you enter the blackness, your lights probing the darkness like a laser. Under the beam of the light animals of rare and unusual shape are mesmerised by the light. Divers can handle interesting fish that are unapproachable by day. In twenty-five years of night diving there has never been recorded an attack at night, by any sea creature so there is no extra cause for concern. The excitement of diving at night, the thrill of the unexpected is always there and once divers are over the unrealistic fear of the dark the experience becomes one of the most exciting in diving.

The marine research diver course introduces the sea's inhabitants and develops appreciation for conserving the marine environment. On completion of each of the above courses, a

Night diving is a popular recreational activity. Here students are going through a predive check before commencing a night dive. Mounted on the divers' tanks are cylume sticks or chemical glowlights that last up to 8 hours in case the divers' torches go out.

certificate is issued which gives you the required credits towards instructor training should you decide to pursue a career in teaching diving.

Underwater photography is a fascinating hobby. At its highest level it can provide the satisfaction of achieving Oceans Underwater Photographer of the Year. Australia's top underwater photographers exhibit their competition entries at the festival, which is held annually in Sydney and Melbourne, and attracts crowds of up to 3000 people at each venue. The South Pacific Underwater Photographer of the Year award is also presented annually and attracts many entries as it is a prestigious award. Underwater photography and photo-journalism is a lucrative field and with three underwater magazines in Australia, there is a market reward too. Beginners in underwater photo schools can be professionals in other areas of photography. The first hour of training teaches basic photographic techniques and after that, pupils learn how to handle the equipment and themselves underwater. Setting up the equipment correctly to prevent flooding and light and colour balance underwater are unique problems for the underwater photographer. Optical considerations and the different lenses are factors that require special consideration in the underwater environment. How to judge exposure when using strobes and light meters is another technique covered in the course. Video is an expanding area in underwater photography, now that compact models, which can be installed in underwater housing, are available . The manufacturer, JVC, produces several models, and underwater camera housings by Ikalite, See and Sea and Seatite are available for video cameras. Lighting for video is much simpler than for professional movies because less powerful lights are adequate and are already available. The Swiss firm Sub-

Photographer Kevin Deacon with an underwater movie camera and lights, photographs the sea whips on a coral reef.

atec have advanced models of underwater lighting for video. The advantage of video is that the entire dive can be recorded and then shared on the dive boat with your companions. Also the inexpensive running costs — you can erase and refilm on video without additional expense — make video an alternative to traditional forms of underwater photography. Participants in photography courses have the option of hiring equipment and learning what type best suits their needs before they purchase, or they can learn to use their own equipment.

The Nikonos underwater camera was a pioneer underwater camera and is still a suitable choice for the beginner or the professional. It has a range of optional extras that extend its capabilities for special performance, yet it is relatively inexpensive. If you intend only to use the Nikonos underwater, the 28mm lens will best suit your purposes, but if you also intend to use it for general above-water photography the 35mm lens is the best to buy. Macro tubes and close up lenses enable you to get closer to the subject and photograph micro life when conditions are adverse. Nikonos manufactures a strobe light with an automatic point-and-shoot system that

The Nikonos Five underwater camera and twin SB102 Nikonos strobe lighting units are an excellent choice for beginner or professional.

The underwater photographer with his dome port camera and lights photographs a sea star on a sponge.

is expensive to buy in Australia, but many other manufacturers offer mini, intermediate, and powerful strobes, some to match certain lenses and some of which are more readily serviced than others. The Nikonos is light and compact for general underwater photography and has a range of other cameras for special purposes. Housing a camera you already own is usually more expensive than buying a Nikonos, but there are many excellent underwater housings available. It is important to ensure that the above-water camera has a removable top prism so that an action viewfinder can be fitted which enables focusing even when the diver is wearing a mask and the camera is in a housing. It is surprising how few advanced cameras have this feature, which is essential for successful underwater photography.

Special cameras are needed to photograph above and below the water surface, simultaneously. For this purpose a Canon F series camera in a Seatite housing, with special dome port and split level focus lens, is excellent. The telephoto capacity of the Nikon F series camera in a Tussey aluminium housing and the single lens reflex facility are ideal for wildlife photography. For super wide angle photography the Aquatica is ideal and this housing, with a Nikon F series camera, is used by many professional underwater photographers. Subatec is the main strobe lighting that Kevin uses for wide angle work. It is strong, compact and reliable with four power settings and it has a modelling light which illuminates the subject for focusing and strobe alignment. AquaSea makes a range of strobes in Australia which have most features that both amateur and professional photographers require. They are lightweight, inexpensive and their performance is exceptionally good. Service can be a problem with some strobes but AquaSea strobe lighting is made in Australia, and can be readily serviced here. Kevin uses mini strobes by AquaSea for close up and macro photography, sometimes using two lights in a tandem arrangement.

Cave diving is another specialist area of diving where training is essential to increase the diver's skills. A number of fatalities in the freshwater caves of Mount Gambier's limestone region in South Australia in the 1960s and early 1970s led to restrictions on all divers entering the caves. Cave diver training is a prior requirement to diving in the caves and applies to all divers regardless of their expertise. The Cave Divers Association of Australia issues category cards. Category one dives are easy caves where the water surface is directly overhead at all times and risk of siltation is minimal. Category two sinkholes are big and often deep while category three sinkholes are small, often deep and easy to become lost in, and require the highest level of training. The imposition and enforcement of these restrictions in 1973 successfully eliminated deaths in the caves for ten years. Unfortunately two divers recently drowned in the caves. The use of category cards is enforced by landowners who refuse to permit divers to enter the sinkholes under their jurisdiction, unless they hold the appropriate certification.

Cave divers are taught to use safety lines and to plan their dives so that they can explore the passages safely. One safety rule, so that divers don't use all their air and have none for emergencies, is known as the one third rule, that is, one third of their air to go in, one third to come out again and one third reserved for emergencies. Safety lines are lifelines to cave divers. The lead diver unreels his line and his companion clips himself to the line. Obstructions or bends in the cave passages, necessitate tie-offs. Lights are essential as many of the sinkholes have very small access holes, limiting the natural light available. Sometimes rope ladders may be needed to reach the water and abseiling skills help divers get themselves and their equipment to the water.

Cave diving is good training for diving in dark dirty waters as commercial divers often do. The hard hat diving gear worn by Harold Chadwick in his days as a commercial diver in Sydney has been replaced by the Kirby Morgan equipment, a fibreglass helmet designed for space travel and now modified for diving. The changes in equipment have not changed the fact that commercial divers still perform difficult and often dangerous work, but safety has certainly improved, as has the financial reward. Checking the Snowy Mountains Scheme tunnels at Burrinjuck Dam in 1953, Harold Chadwick worked for seven days in fifty-six metres of water for the grand sum of £20 a day. Today divers engaged in similar work are paid $40 000 per annum. Harold trained in the navy and worked with and trained many of the commercial divers in Sydney. Later some of them sought lucrative work in abalone diving in Bass Strait or on oil rigs. He was the first contractor to work on the Sydney Opera House. Another early Sydney diver, Johnny Johnson, salvaged £2.5 million in gold from the Niagara shipwreck in 1941, which would be quite a feat even with today's equipment.

Commercial diving has a long history in Australia beginning with the pearl diving operations based in Broome and on Thursday Island. The graveyard at Broome bears witness to the lack of knowledge of diving safety, the 'bends' took many victims in those days. In 1875 Bucquoy proposed the 'bubble theory' as the principal cause of caisson disease or the 'bends': after breathing nitrogen at pressure, when that pressure is relieved, the nitrogen takes some time to come out of the blood and bubbles may form causing paralysis. In 1869 a recommended decompression rate of one minute per metre was suggested and in 1875 the first recompression chamber was built to treat workers on the Brooklyn Bridge Project in the United States of America. In 1878 the first compression and decompression schedules were published but it was 1907 before they were in general use. Today's British Admiralty Tables derive from those tables computed by the eminent physiologist Professor J. S. Haldane. By following the decompression tables we can now dive safely but the evolution of safe diving practices has taken almost 100 years.

Above: Kevin Deacon is about to enter a large ocean cave at Lord Howe Island for a long photographic dive.

Below: Diver explores a cave almost overgrown with coral on Australia's Great Barrier Reef.

Commercial diving involves working on the sea floor, or underwater structures on the offshore oil rigs, or in the harbours, or even the rivers and dams. Today commercial divers are doing similar work to that done 100 years ago. However, in that time, the equipment and techniques have improved so that now commercial divers wear the Kirby Morgan mixed gas equipment, rather than the old copper helmets. They breathe a mixture of oxygen and helium that enables them to work at much greater depths.

In the early days of scuba diving members of

The diver descends to a coral cave on the edge of the reef from an 8 m dive catamaran.

As the divers go down the anchor, the sunrays silhouette their progress.

Commercial diver training in the use of underwater cutting equipment. The Oxyarc uses oxygen and electricity to generate sufficient heat to cut steel. An organisation in Morwell, Melbourne, trains all divers for offshore oil rig diving and harbour activities. The Kirby Morgan headpiece allows the diver to communicate as a telephone connects to the surface with the umbilical cord which supplies the diver with air. All commercial divers belong to PDAA, the Professional Divers Association of Australia, a union which sets safety standards, wages and conditions.

the public had rather unrealistic expectations of divers. People wanted a diver to find items they had lost overboard from boats, even from the Manly ferry. Kevin and I used to receive endless requests for jewellery,watches,rings and dentures. Despite warnings that it was unlikely that we would locate them, we were sent off to search the harbour. I remember one customer who had lost a ring at Quarantine Beach near Manly. She had lost it from a moving yacht and had managed to pinpoint the spot from the moorings nearby. With severe reservations, I went down and began a search pattern. Before very long I spotted the ring sitting upright on the sandy bottom. I don't know who was more surprised — me when I found it, or she when I popped up wearing the ring on my little finger.

Many of the jobs done by commercial harbour divers involve routine work such as laying Telecom cables across harbours and rivers, placing water pipelines across bays and, as Kevin describes, this can be quite difficult:

> One job that stands out in my mind was laying a pipeline in the Hawkesbury River. For this type of job, divers usually use hookah equipment, a compressor on the surface pumping air to the diver down a hose. The job stands out in my mind because of the strong current that raged. I had to haul myself hand over hand down a rope to the trench I was digging with the suction dredges and lash myself to the pipe I was installing. The suction dredge digs a hole for the pipe to be buried into the bottom. The Hawkesbury was full of jellyfish that kept bumping into me as I worked. On one occasion, a surfacing diver lost his grip on the line and was swept away to the full length of the hookah hose. This had happened previously, but on this occasion the hose parted at one of the fittings and the diver was swept down river at a speed of five knots. It took some time to get the boat tender into the water and the diver was about two kilometres away when picked up. Each day we would drive to the site to be picked up near the Hawkesbury Bridge and taken to the site. While we were waiting, the inmates of the mental hospital near where we met the dive boat approached. They were fishing without fishing lines, and shooting at the ducks without guns and they involved the divers in these activities. It became part of the day's routine and after a few months we began to wonder who was more sane: those at the wharf or the divers.

Salvage is another aspect of commercial diving work, one which sports divers picture as the most common aspect of commercial diving but which is rather infrequent. The vessels salvaged range from huge ships to small runabouts and it is not always the size of the vessel which determines the difficulty of the task as Kevin explains:

> I was employed to salvage a steel motorboat. It had sunk at a marina in Pittwater, north of Sydney. I had strapped flotation drums to the vessel. These are filled with water and taken to the bottom, then the water inside is replaced by air under pressure,

> giving them buoyancy. The drums are roped together and to the vessel before they are filled with air. I tied all the drums I had with me to the boat and filled them with air. I could lift the stern of the ten-metre vessel in my hands, like a superhero, because it was almost weightless. The air-filled drums had attained neutral buoyancy. As I lifted the boat it sped rapidly to the surface. I came up, breaking the surface in time to see the boat sink again. I wondered why, but went down to the bottom where the boat now lay and refilled the drums. Once again, as I surfaced, the boat was on its way down. I realised, finally, that when the boat popped above the surface and settled back, some air in the drums was squeezed out and it lost the critical buoyancy that was needed to hold it on the surface long enough for the diver's assistant or tender, to secure it. The problem was solved by tying ropes to the boat's bow and stern, port and starboard. Assistants were instructed to keep the lines taut and lash them to the bollards on the wharf as the boat rose to the surface. On the third attempt, we had it on the surface and the pumps were started to empty the vessel.

The procedures for different diving jobs often have to be worked out on site and experience in similar situations gives the commercial diver a considerable advantage. However, situations sometimes arise for which the diver is not prepared and these can be dangerous even for experienced commercial divers. One such incident involved a commercial diver scrubbing a hull on a large ocean-going liner. When the big ocean vessels come into Sydney after a long voyage, they have accumulated marine growth on the hull. The drag from the barnacles and weed reduces their speed considerably. The reduction in speed on longer voyages means more fuel and a slower passage, so substantial savings can be made by cleaning the growth from the hull while the ship is in port. Divers originally used scrapers and ranged all along the ship's hull taking a few days to remove the growth. As procedures were refined, they began to use a water jet gun. The water jet has a pressure of 3000 to 4000 psi and ejects from both the front and rear of the gun. The water must come out in both directions to attain equilibrium, so that the diver is not thrown about, as he could not control such a force underwater, if it was not balanced. The water jets blast the weed and barnacles from the hull. The first time a water jet was used on a commercial diving job it almost resulted in a fatal accident. The manager of one of the largest commercial diving operations in Sydney, was working on the job. He swam down to relieve a diver on a shift changeover. The diver operating the water jet swung around in the poor visibility and accidentally swept the back jet of the gun across the relieving diver's shoulder. It cut straight through the wetsuit and into the shoulder narrowly missing the neck. The accident victim required twelve stitches, but the jet could easily have cut through the jugular vein which would have killed him. The procedure was established to avoid a similar situation: the relieving diver would indicate on the operator diver's line that he was coming down; the operator would aim the gun at the hull and the relieving diver would take over, giving the operator time to leave the water before beginning work. Procedures to avoid accidents have been established only after fatal accidents in some cases. One such incident involved two divers who entered a ship's hull to release a valve that was blocked and was retaining water in the hold. The divers entered the water in the hold, released the stuck valve and were leaving the hold on a ladder, out of the water, when they were overcome by toxic fumes, collapsed and died. Divers now wear breathing equipment in any confined space where there is danger from toxic fumes.

Commercial divers can have some unusual experiences with marine animals when they are working in the dark, and often dirty, harbour waters, as Jeff Deacon discovered:

> I was working on a sunken launch in Double Bay, a Sydney suburb. The water visibility was about a metre. I was kneeling on the bottom attaching drums to the vessel in order to raise it to the surface. The bottom was clothed in drifting seaweed and I felt something wriggling between my knees. I thought it was my companion passing me a rope and felt for it, but it bit me. I rapidly withdrew my hand and hanging off the palm was a dogshark. It's a small bottom-dwelling shark with small teeth which is not usually a hazard to divers. I dragged it off my hand and headed for the surface.

Jeff experienced another nasty incident while working on a wreck. He and Kevin had been exploring the Malabar wreck, off the Sydney suburb of Malabar. Jeff noticed something protruding from the sea floor and attempted to remove it forcibly. As the object came with a rush Jeff bumped into a sharp projection which pierced his elbow to the bone. This wreck is in a sewerage outfall area and the wound became

Chris Deacon strokes a potato cod at the Cod Hole, Ribbon Reef Number 10, near Lizard Island.

badly infected and kept him out of the water for months.

Professional divers in Australia belong to the PDAA, the Professional Divers Association of Australia. With the increasing need for commercial diver training to supply divers to work on the Bass Strait oil fields, the Association in conjunction with International Seawork, founded the Underwater Training Centre in 1976. The National Safety Council of Australia took over the Underwater Training Centre in 1983 and relocated to Morwell, Victoria, in close proximity to the oil fields.

Basic air diving and underwater work is a twelve-week course at the centre and includes underwater cutting and welding, underwater explosives and construction techniques. The bell diving training course has a prerequisite of twelve months commercial diving experience over a range of depths to fifty metres. The seven-week course includes mixed gas saturation diving and deck decompression. A range of courses of shorter duration are also available. While commercial diving is a very rewarding occupation, there are many other opportunities to earn a living from diving and these include aspects such as underwater photography and cinematography and diver training, retail marketing, marine biology or other specialities such as photo journalism.

Divers often pursue their shore-based interests underwater and this, too, can lead to career opportunities. Painting and graphic arts are unlikely activities but artists are now venturing with oils and brushes, easel and canvas, to record their impression of the underwater scene right where it happens. Marineland is an ideal place to begin: there is plenty of exotic life to record on canvas. Whether you paint in oils or use watercolour or poster techniques, the ocean, its inhabitants and its scenic beauty, provide inspiration that few have attempted to express. Watercolour, too, is a medium which has its underwater artists. Painting underwater or at home in the studio, conveys the wonders of this fragile environment to a wider audience, those who have not yet ventured down under.

The painting by Yak Carrasco (above) of a wentletrap or ladder shell displays a stylised rendition using graphic design techniques.

Above: Brian Maloney captures his subject, a grey nurse shark, underwater in Marineland, Manly.

Below: This oil painting, 'Comes a Predator' (183 × 120 cm) by underwater artist Brian Maloney, can be seen in the Sydney Hilton Art Gallery.

Above: Black lip pearlshell medallions are the feature of this Rosalba Nelson design worn by model Alana Scowen at Warriewood, Sydney

One such artist is Yuk Carrusco, a Colombian, who migrated to Australia in 1981. He learned to record his impressions of the ocean using the very essence of the sea, salt crystals and watercolours, evolving the special techniques to capture his subjects. Through his art form, people who may never have been underwater become more aware of the fragile and beautiful environment and the need to conserve it. His strange hallucinatory works are a product of his Colombian ancestry and his adaptation to his Australian home and the seascapes he loves to paint reflect this artistic growth.

A different art form that captures the beauty of the sea is jewellery making. Many pieces by Rosalba Nelson are inspired by the beauty of a shell or the shape of a piece of coral; the inspiration for others comes from human destruction of the natural world. Rosalba sees her coffee table pieces in an environmental context. We have destroyed many of the sea's beauties by fishing, pollution and destruction. Even in the death of a starfish or shell, there is a wild beauty that is reflected in the perspex and leather shell pieces she designs to express environmental conscience. Some of the neckpieces have bands filled with water — the shell as it was living in the sea. Others are finely beaded or quiver with life and movement.

Already a well known artist before he ventured down under Brian Maloney works in oils. He has exhibited in Sydney's Holdsworth Galleries, Centrepoint, and the Studio Gallery and is currently writing a book about how pioneers perceived the Australian environment, an experience he has enjoyed in his own pioneering artistic endeavours. Brian uses Manly Marineland in Sydney as an underwater studio where he produces surrealistic paintings as well as collecting animals for the aquarium when he has the opportunity to study them in their natural habitat.

Sometimes divers like to take their diving interest home into their living room. Keeping marine life in an aquarium is challenging and you come to know the fish intimately. Some aquarists keep fish for many years — I know of one diver whose tropical angel fish collected on holiday was still keeping its owner fascinated seven years later. Aquariums can be any size from a small plastic bowl to a giant tank, recreating a section of ocean. One in Auckland, New Zealand, designed by the late Kelly Tarlton gives non-divers an ocean experience as they pass through a tunnel under the aquarium, without even getting wet. Many marine creatures can be kept in an aquarium, even live coral or algae, but it is essential to have good filtration to remove urine and other fish wastes from the water, a subsand filter or bionic filter is preferable. Marine plants need special fluorescent light. Marine fish will feed on coral or algae in the tank but need supplementary food such as brine shrimp, tubifex worms, fresh shrimp and lean beef. A great deal of skill is required to keep marine fish in aquaria and diving friends will be greatly surprised by your depth of knowledge of fish habits.

Collecting is another favourite diving pastime. The old jetties and picnic sites around harbours and foreshores are interesting sites for digging up history. Finding a glass stoppered bottle or stone ginger beer bottle are the rewards of digging in the muddy bottom. Boat anchors often disturb the seafloor exposing all sorts of rejects from the past. Collecting coral or shells, once a popular activity is now environmentally disastrous. The trumpet shell or triton collected over many years on the Great Barrier Reef is thought to play a part in controlling the numbers of crown-of-thorns starfish. The triton eats the larvae of the starfish, thus reducing the number that reach adulthood, which is when they feed on coral reefs. Since we do not know how important different predators are on the starfish population, or any other potentially nuisance species, it is best to leave shells and other marine life where you find them.

Divers like to harvest the fruits of the sea. Abalone is a special delicacy of the southern areas of Australia. Diving to the bottom in three or four metres of water and looking for 'abs' under rock ledges is an activity beginners and old hands alike can enjoy. In my early diving days, searching for seafood was my favourite diving activity. After a few dives I'd find what I was looking for: a ledge with a few abalone clinging onto the sides of the rock. I would prise off a few and head for the surface for air. After a few more dives I'd have enough for a meal. Spearfishing, once a very popular diving activity, has declined in importance as conservation has gained momentum.

Sea urchins, mussels, scallops, lobsters, fish — seafood is a gourmet's delight. The thrill of capturing it and the joy and anticipation of a flavoursome meal amongst friends is one of the many aspects of diving. Whether your tastes are for exploration or collecting, for watching or recording your impressions with camera or paintbrush, the underwater world has much to offer. We have seen in this chapter how diving can be a rewarding pastime and how it can offer career opportunities. In the following chapters, we look at different parts of Australia, from the northern coral seas to the cold waters of southern Tasmania.

Divers are holding their cameras high in triumph at the end of their photo course. Kevin Deacon trains underwater photographers at courses organised around Australia.

An underwater photographer in action, photographing the pink soft coral with its yellow polyps. Three flash units have been used to provide the lighting for the photo. Chris's flash is responding to the flashes on the photographer's camera.

DIVING CORAL SEAS

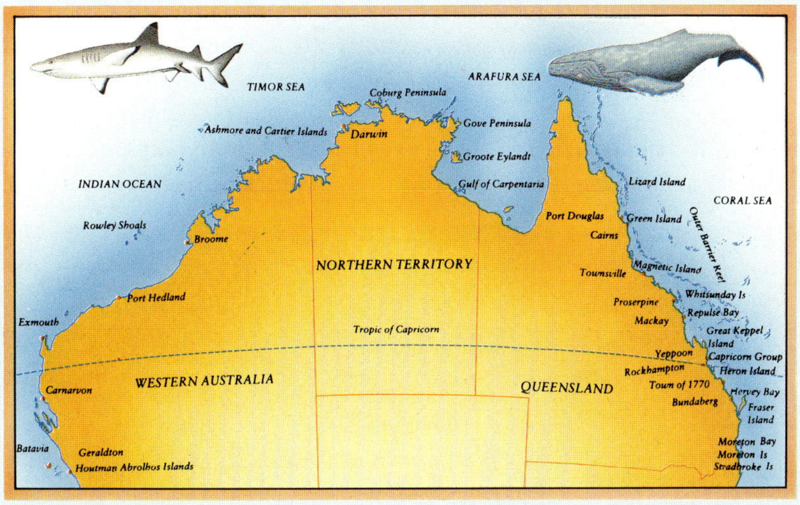

Like underwater jungles, coral reefs flourish in the warm tropical seas of Australia's north with its abundant sunshine and oxygen from the constant wave action over the coral heads. Though it was originally formed between twenty-five and five million years ago, the Great Barrier Reef had its major growth around two million years ago, the time when the whole of northern Australia became truly tropical.

Although the reasons for the sudden spurt of growth of coral reefs around Australia are not known, scientists believe a particularly rapid rise in sea level washed the surface of the continental shelf of mud; this allowed coral larvae to settle and grow. The long period of stability since then has enabled thousands of marine creatures to evolve and thrive in the colourful coral gardens. Today, the main differences existing between the coral reefs of eastern and western Australia are that firstly, the western reefs, though comparable in extent to the Great Barrier Reef, contain far fewer individual reefs and secondly, the inshore fringing reefs of the west support animal communities which reflect the coast's lack of rainfall and major coastal rivers. Because there is so little nutrient-rich freshwater input, many of the animals which are found hugging the fringing reefs of eastern Australia are unable to survive the west.

The large diversity and numbers of marine creatures — scientists recognise over 30 000 different animal species living in the waters of the Great Barrier Reef — may explain the splendour and brilliant colours which characterise coral reef life. Most creatures are strongly marked and richly hued. Many parade violent contrasts — yellow and black, white and orange, green and pink. Exactly why they are so vividly adorned is not fully understood, although protection from predators and mating habits — both elements of adaptation and survival — are likely to be involved. Whatever the reasons, the jewel-like colours of reef animals undoubtedly remain one of the most compelling attractions for divers of coral seas.

"My first impression of diving coral seas was the crystal-clear blue water. The calm surface with turquoise reefs shimmering below enticed me in and wrapped around me. I felt that I was falling like a leaf, onto the coral reef below. But the buoyancy of the water supported me, its clarity was so deceptive that it had tricked me into thinking there was nothing between me and the bottom. The brilliant coloured fish darted into a coral clump and hid from me. The black and white humbug fish faced me defending his territory. I swam towards a coral bommie, an isolated block of coral — a column rising from the seafloor to the surface, like an apartment block in the sea, with inhabitants at every level. A turtle swam up to see what I was looking at and before I could grab his back for a ride, he flipped into the distance. A small reef shark swam past ignoring me as he whipped into the shallows above the reef. The time passed so rapidly as I watched the myriad inhabitants of the reef go about their daily routine, that the tide changed and I had a long swim back to the boat."

The little clown fish is tucked away for protection among the sea anemone's tentacles. Other fish attempting to eat the clown fish would be stung by them.

The biggest coral region in Australian waters lies off the Queensland coast, but there are also coral reefs off Western Australia, the Northern Territory and New South Wales. The Great Barrier Reef is one of the most densely populated areas on earth. It is an enormous coral reef consisting of thousands of separate reefs, some rising 200 metres from the ocean floor. Every square metre consists of thousands of coral animals. Most of the food comes from the rainforest, swamps and deep ocean water which are the source of nutrients carried by the ocean currents to the reef. The living creatures of the reef capture food in the water, like a giant filter. Nothing is wasted. The reef is built by coral polyps which have been on earth for 500 million years. This reef world has remained intact over millions of years, while whole species of larger animals have been wiped out. The sea has protected the animals from changes that have affected the land.

Nature demands changes and the reef is used to them. They have always been part of the natural processes. Cyclones tear up huge sections of the reef. The last Ice Age saw the reef disappear to the edge of the continental shelf, yet it recovered and began to grow on the old formations where the water depth and temperature were suited to coral growth. Coral reefs have withstood atomic blasts. The atolls used for testing have grown back in thirty years. Outside the reef, the water is deep and channels penetrate the reef so that it is a series of separate reefs interspersed with deep channels. The passages through the reef were difficult to find and many ships were wrecked on the coral. Even with modern equipment travelling on the reef still requires caution.

One of the oldest races on earth, the Aboriginal people, have lived on the coastal lands and fished the reef, accepting the land and waters in trust for over 40 000 years. Their legends tell of the history and origin of the land. One legend relates to the water ways of the rainforest behind Cairns.

> In the forest lived a tribe including a beautiful girl named Ouruna and a well respected elder named Eurana. The chief was pleased with them and gave Ouruna's hand in marriage to Eurana. They lived happily for a time until a wandering tribe entered the area. A handsome man named Deaga fell in love with Ouruna and the two decided to run away to the headwaters of the river. They lived here until the two tribes came searching for them. The tribes wanted to take Ouruna back to her husband. They grabbed Deaga and took him away and were taking Ouruna back to her husband. When she knew she would never see Deaga again, she threw herself into the waters of the Devil's Pool, and drowned. To this day her spirit is there. She calls to any young wandering men, not in her tribe, and they fall in and drown. Over recent years seven young men have drowned in the Devil's Pool.

Fringing reef and inner lagoon of Lady Musgrave Island. The windward side has surf breaking on the reef and the leeward side has a protected lagoon.

Some of the legends go back to a time when the mountains were active volcanoes.

The reef is a living system composed of ani-

During the dry season in winter when water levels are low, swimming in the rock pools is fun.

mals and plants which interrelate. Most of the plant life is microscopic and is often found in the tissues of animals, such as coral, which it provides with food. Human time scales and geographic areas are meaningless on the reef. Like the Great Wall of China, the Great Barrier Reef is visible from the moon. Since Europeans first saw it they have been afraid of its dangers. Countless ships have been wrecked along it, often after long voyages of thousands of kilometres and within sight of their destination.

Marlin have attracted many overseas visitors to Cairns, where marlin fishing has become the spearhead of a newly developing tourist industry. For three months of the year twenty-five boats and a handful of mother ships fish the offshore reefs in a stretch of water 200 kilometres long. Every marlin caught is tagged and released, except an angler's first or biggest fish. Most visitors to Cairns come from down south or from overseas. Cairns is the centre of the tourist industry in northern Queensland. More than half the population is involved in tourism, and more people are self-employed there than in any other Australian city. Cairns has a very cosmopolitan population, and the area has many attractions for the visitor: from the rainforest and the reef to the more sophisticated lifestyle.

The rainforest is the richest biotic community on land. Rainforest originally lined the coast adjacent to the reef, which is the richest in the sea. The two ecosystems form interrelationships that are important to both communities. The Ice Ages and changes in sea levels have shifted the communities but both have been retained through time. Aboriginal people lived in the rainforest of north Queensland, adapting their lifestyles to the seasonal abundance of fruits and foods. The density of plant life amazed Joseph Banks, the first botanist to see the rainforest, and frightened the Europeans who followed. Today forty per cent of the world's medicines come from rainforests, many from the fungi and decomposers which live within the forest. For thousands of years a complex relationship between the rainforest and the reef has existed. As well as controlling the run-off from the heavy rainfall, the compost formed on the forest floor is food for the river systems. The fungi within the compost gives antibiotic properties to the rivers which purify the streams and may prevent disease in the mangroves. There exists a complex and stable system that has evolved with many interrelationships over millions of years. Ninety-five per cent of Australia's rainforests have now been exploited for agricultural and other land uses. On a worldwide basis only half the forests that existed in 1950 are still standing. We don't need the space for development in Australia, yet we are still destroying the remaining stands of rainforest. Some of the trees may take 1000 years to mature, a time span that is awesome when we are accustomed to think only in terms of our own lifespan.

The river systems that feed the reef are an important part of its ecosystem. The reef provides shelter below the water and the cays provide a breeding ground above the water for birds. The interaction of the many facets provides the key to understanding the system. Mangroves provide the productivity in the estuaries for the prawns and birds to subsist, and not only in the estuary, but offshore as well. The mangroves are one of the most productive areas in the world comparable to cornfields in productivity. Bacteria break down the leaves of the mangroves and the release of nutrients provides the basis of the food chain for barramundi, our premier food fish, and mangrove crabs, as well as the many other animals. The twenty-eight different types of mangroves found in Australia are important to the whole system. The seeds are carried to a favourable spot and germinate. A true evergreen forest develops, adapted to the high salt of the sea. If you taste the undersurface of the mangrove leaves they have a salty taste, because salt is exuded through the pores in the leaves. When a cyclone strips the leaves, the mangroves die because they cannot get rid of the salt. Severe storms, although they destroy sections of forest, help to maintain diversity by shifting the balance of different types of trees that are present.

On the coral cays, birds provide the guano from their droppings. The seeds washed here, or dropped by the birds, take root and hold the sand together as it builds up. Eventually a vegetated island is formed. Green Island is such a cay as is Lady Elliot Island, where guano was mined for many years. The cays extend at intervals along the reef, interspersed with high islands — the tops of hills, which were once dry land, but are now surrounded by sea, many with fringing coral reefs.

Green Island is a coral cay with the first

Chris Deacon preparing Sea-Tite 650 watt movie lights powered by a surface generator and cables which are used to provide light and colour balance for underwater filming. Kevin and Chris were filming on Green Island Reef for a documentary on the Great Barrier Reef. There are many giant clams on Green Island Reef although the reef is shallow and does not have many hard corals.

underwater observatory in the world. It was on Green Island that the threat from the crown-of-thorns starfish was first recognised. An aquarium on Green Island has caught and maintained fish and turtles for over ten years. Some of them were young animals when caught but are now huge. A loggerhead turtle has grown to over a metre across. While most visitors visit Green Island, few have the opportunity to explore the eleven million hectares of reef.

Giant clams are one of the fascinating sights at Green Island and the northern reefs. There are seven species in the world and all are found on the Great Barrier Reef. On pristine reefs they may be clumped together in groups of the same species.

Giant clams have an interesting symbiotic relationship with the microscopic algae in their tissues. These algae are zooxanthellae, similar to those found in corals. The algae live in the clam receiving carbon dioxide and nutrients while providing the clam with simple sugars, glycerol and other products from their photosynthesis. They may supply up to fifty per cent of the clam's food, the zooxanthellae are free within the clam's blood system which is unlike in the corals where they are embedded in the cells. The algal cells are not transferred in the eggs, so the relationship must be re-established each generation. The variety of colours, associated with the mantles of giant clams, is partly due to the algal cells and partly to the light refracting iridocytes, cells that refract light to produce mainly blues and greens. The iridocytes may protect the clam from ultra violet light. There are various mantle colours and patterns which can be seen when the

Chris examines a rather small giant clam which is amongst the coral. As it grows it pushes aside the coral surrounding it and as its valves open and close it crushes the coral in the immediate vicinity of the shells.

The magnificent filigree pattern of the gorgonia can be seen as Chris examines the orange gorgonia. The name gorgonia comes from the Latin meaning 'many headed creature' and refers to the colonial nature of the animals.

clams' shells are ajar. The eyes of giant clams are on the exposed mantle along the edge of the shell. When a shadow passes over the eyes, the clam responds by closing the shells, but not so rapidly to trap a diver as legend claims. The famous Pearl of Allah, a 6.5 kilogram white pearl was found in a giant clam that was said to have taken the life of a diver who discovered it.

On the Great Barrier Reef about thirty foreign clam vessels have been captured while poaching in the past twelve years. The Commonwealth Fisheries charter private vessels to catch boats spotted from the air. The navy also apprehends foreign vessels fishing in our waters. The clam poachers kill the animals for their adductor muscle, the one which controls the opening and closing of the shells. Over the last twenty years, millions of clams have been taken from the reef. Only about ten per cent of the clam is used, the rest is left, in a very wasteful fishery. Such a systematic harvest of one animal is potentially dangerous. Even the risk of losing their boats does not deter the fishermen.

Pacific Islanders and Papua New Guineans have been farming clams on a subsistence basis for years. They carry small clams to a reef adjacent to their villages and allow the clams to grow to a size large enough for a feast. Fisheries workers in various Pacific nations have been working towards the cultivation of giant clams, to re-seed the overharvested reefs throughout the Pacific and Indian Oceans. Giant clams are shellfish which could be farmed on a large scale. They reach commercial size in two years. They are hermaphrodites, that is they produce sperm, which is spawned first, then several hours later the eggs are released. Natural spawning occurs around new and full moon in December. A powerful jet of water with the sex cells shoots to the surface with enough force to be visible at the surface. The close proximity of other spawning adults increases the success rate of their reproduction. High densities of more than 100 clams per hectare (or more than forty per acre) are

The Spanish dancer is a shell-less mollusc which grows up to 50 cm and undulates in the water giving it its name. The head and rhinophores or tentacles are visible. Both the rhinophores and the gills on the rear end can be withdrawn.

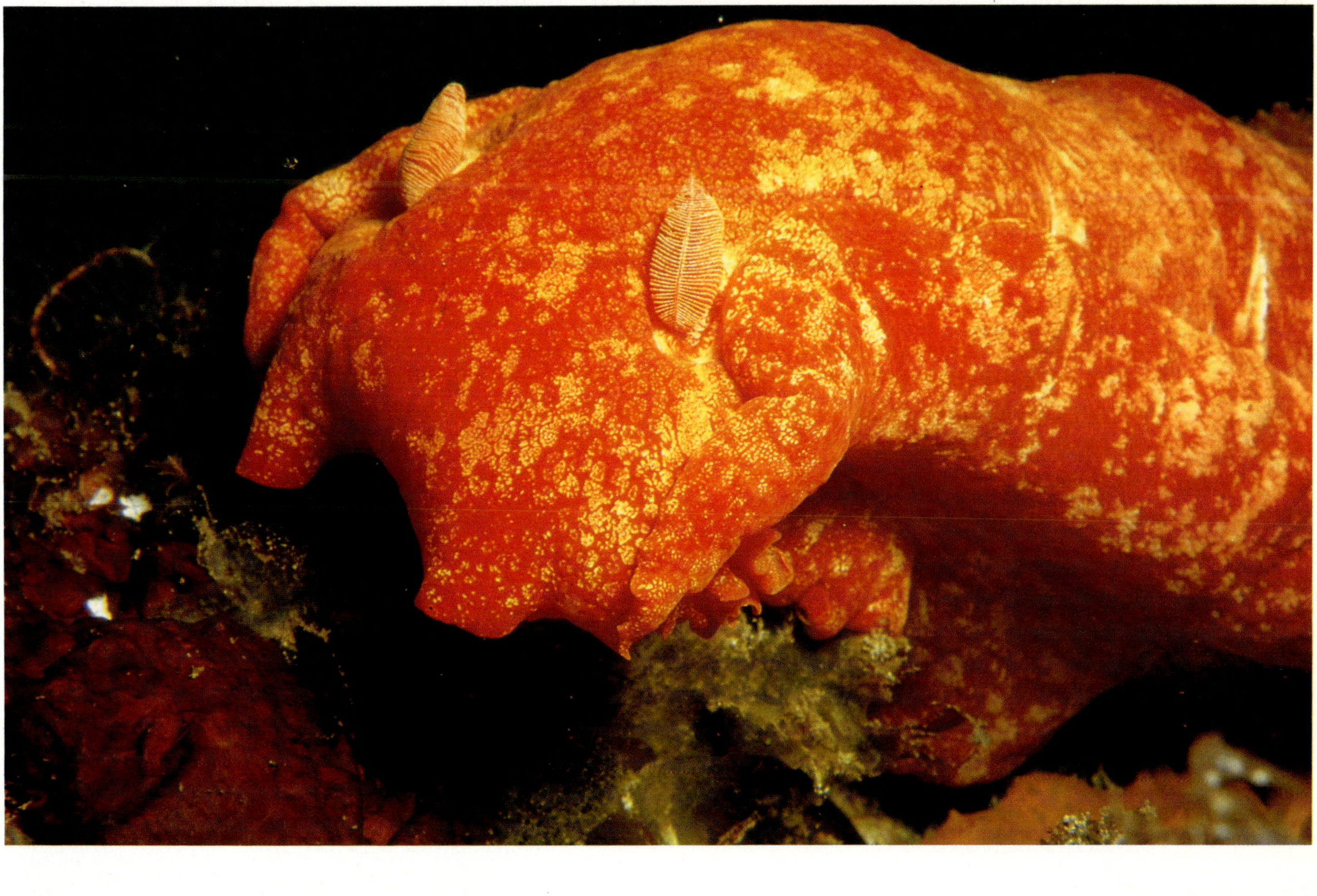

Chris Deacon has encountered a large tropical cuttlefish which is a different species to the larger giant Australian cuttlefish and is lighter in colour, although it is still able to flash some brilliant colours when aroused.

known. Where such high densities occur, they are clumped together by species. At such high numbers, the shell edges may cut through adjacent clams. The clams' eggs move in the plankton for seven to sixteen days before they settle on the bottom. This is the stage when many are lost. A hatchery on the Great Barrier Reef has been proposed to raise the larvae. The clams raised on a stick and tray culture system similar to that used for Sydney rock oysters could be farmed and used to re-seed areas that have been overharvested.

Night diving is one of the fascinating experiences on the reef. All the fish are asleep except the night feeders. Divers with lights can pick up and handle fish. The parrot fish wraps itself in a cocoon, probably as a protection from morays which hunt at night. If the parrot fish is blinded by the light it will allow the diver to handle it, but if it wakes suddenly it will explode out of the cocoon. The cocoon around the parrot fish protects it by preventing the release of its smell in the water. If the moray touches the cocoon the movement acts as an early warning system, and awakens the fish. The reef at night is spellbinding: all the corals are out feeding and it is possible to see many creatures not normally seen in daylight.

Fish have evolved an amazing range of habits for survival. The potato cod, like a number of other fish, changes sex as it grows. Most of the reef dwellers are territorial and small cod have little chance of competing for living space, so they start life as females. The males gather harems and the small female cod are guaranteed space on the reef. When they become big enough to obtain some space for themselves, they change sex. They take over a territory and gather their own harems. There is a pecking order. These fish have developed a lifestyle to maximise their chances of survival.

At the cod hole off Lizard Island, divers have been feeding the potato cod for years. The cod hole is on the northern end of Ribbon Reef No. 10 and is an indentation in the reef. From the moment the boat anchors in ten metres of water, the cod appear. They are huge fish and the diver entering the water is soon surrounded by them. The fish will grab any food from the diver's hand — taking glove bag and all. A solid bucket is the answer, then the fish will wait until you reach the bottom, and will quickly remove the raw fish food from your hand, sometimes with your glove, which they will spit back. When there are ten to twenty cod, from 100 to 300 pounds, feeding at once and in ten metres' visibility, that's pretty exciting. Once they've been fed the fish will permit the divers to stroke and pat them scratching them under the gills, the fish seeming to enjoy the encounter. There are some old fish that are very comfortable with divers. You can recognise them by the faded grey, others have different patterns and markings. A wave of the glove will bring them over. The cod hole is coral, which is not very profuse. A special protection applies to the cod that are famous worldwide. Ron and Valerie Taylor were instrumental in having the area specifically protected. The spot has many different species of fish, including moray eels, that are now accustomed to being hand fed but it is the huge potato cod with their comic antics and ferocious appetites which make the dive one of the most exciting on the Great Barrier Reef. Most of the charter boats going north from Cairns, visit the cod hole. It is about one and a half hours travel, by boat from Lizard Island.

Cairns is the gateway to the northern Great Barrier Reef. From here you can join one of the boats cruising the northern parts of the Reef, going to Raine Island, Portland, Roads and, Thursday Island, or you can go on day trips to the Reef from Cairns or Port Douglas.

Between Cairns and Gladstone, the Reef resorts and towns attract divers. Townsville has both inshore and offshore areas to explore. The big diving thrill is the wreck of the *Yongala* which is in the middle of nowhere, swept by currents. It is in relatively shallow water. Once on the wreck there is protection from the current. You can swim at an angle along the passageways, and then even penetrate into the hold but this not as interesting as the deck. The wreck is like a giant reef covered in marine growth and you have to fight your way through the schooling fish, turrum and kingfish, while big black rays swim alongside and sweep past you. There is a cod as big as a Volkswagen living on the wreck. A reef very close to the *Yongala* is also excellent diving. Myrmidon Reef, much further out, is the site of another wreck the *Foam*, a blackbirder, which is in the lagoon. The reefs off Townsville are often horseshoe-shaped — very different from the long ribbon reefs off Cairns. Myrmidon Reef, like many others in the area, just breaks the water surface. On the way there dolphins, even killer whales, may be sight-

Kathie Savage hand feeds a large spangled emperor in the lagoon of Lord Howe Island. This fish has been named Albert and has become accustomed to the divers.

At night, fish with large eyes come out to feed, as this fish did when captured on film by Chris Deacon's camera.

A coral garden with staghorn and plate corals growing on the protected inshore section of a reef. These coral gardens are readily accessible to snorkel divers and underwater tourists.

ed. The huge black dorsal fin of the killer whale is the only visible part as he approaches. The white patches on his head can be seen just before he dives beneath the boat.

Manta rays are the gentle giants — rays with a barbless tail. They are found at various places on the reef but in the Capricorn Group, you are almost guaranteed to see them. Lady Elliot Island has a number of mantas that frequent the waters around the lighthouse bommie, in twenty metres of water. Diving at Lady Elliot is relaxed and informal. Shore diving offers the chance to see the mantas on an interesting coral reef just offshore. The mantas are accompanied by remoras which hang beneath, hitchhiking a ride. Mantas swim in different ways, some undulating their flukes in a ripple, while others flap their flukes like wings. To capture the fluke movement requires a very fast shutter speed over 1/2000th of a second to stop the wing tips, which is the same shutter speed that would stop an aeroplane in flight.

Lady Elliot Island has several other interesting sights such as the blowhole — a large cavern which goes into the reef face about ten metres and opens out on the other side of the reef. There are other caves such as Hiro's cave nearby and there are even rope trails accessible from the shore to The Arches and the Shark Pools. There is no restriction on the number of dives, so you plan your own day's diving to take in as many dives as you please. Even if you don't already dive, some resorts such as Lady Elliot conduct diving courses which are introductions to diving. In good weather boat dives from an eight-metre barge go to the reef side of the island. Access is by plane from Bundaberg, a trip that takes thirty-five minutes. From Bundaberg you also drive out to 1770, where Captain Cook first set foot on Queensland soil and from there you can go out on day trips to Lady Musgrave and other islands in the Capricorn Group on the *Captain Cook*.

Heron Island, also in the Capricorn Group, is usually reached from Gladstone, whether by boat or by helicopter. A coral island right on the Reef, Heron is well known to divers in Australia and overseas. The reef around Heron Island has such well known diving locations as The Bommie, where spectacular coral rears to the surface in a column-like formation. It is here that divers have been feeding fish and moray eels for longer than anywhere else on the Reef. The shark cats and other dive boats go out daily and there are plenty of spots to explore on Heron and nearby Wistari Reef. Heron is one of thirty places in the world where green turtles nest and to see them

Above: Mantas are the gentle giants of the ocean. They are an awe inspiring size yet are harmless, lacking a barb on their tail. They attain a size of about 5 m, this one being slightly smaller and accompanied by 2 smaller rays. They used to be called 'devil rays' because they would rub their eloua against the anchor rope, frightening fishermen who happened upon them. They will approach divers and allow them to ride on their back.

Below: Lady Elliot and Lady Musgrave Islands in the Bunker Group have large families of mantas which will approach divers. They are magnificent creatures which will filter feed using their eloua to scoop up plankton. They will approach divers, inviting them to take a ride seeming to enjoy this encounter.

climb laboriously up the beach after dark to dig a hole for their eggs is an experience you can see in only a few places.

Other fabulous spots to dive at Heron are the Hole in the Wall, where a large crevice runs through a bommie as well as the actual Hole in the Wall which is over the reef edge in depths from ten metres to twenty metres. Blue Pools are east of the Hole in the Wall and cut back into the reef edge. The sandy floor is created in front of your eyes by parrot fish picking up and grinding down coral rubble for food. Then there is North Bommie, or Gorgonia Hole, or the Canyons or the Giant Grouper at Pan's Point. All round Heron's Reef fish are unafraid, and the island is also a haven for bird life, with mutton birds burrowing in the sand beneath pandanus palms, terns nesting on the ground and doves in the casuarina trees.

Hayman Island in the Whitsunday Group is one of the high islands on the Great Barrier Reef — a hilltop now surrounded by sea with fringing coral reefs in some of its bays. Dolphin Point and Blue Pearl Bay are popular sites or divers may visit nearby Hook Island, the largest in the group, which also has the most protected anchorages for visiting yachts and cruisers.

On one of Kevin's early dives in the Whitsunday Islands, he met with some big game. Visibility was not good: ten metres was about the average. Surprisingly this was an advantage, as Kevin explains:

> I was exploring the coral reef when I turned a corner and stumbled on a huge Queensland grouper. I had heard that they grow up to three metres in size, but this was the first time I had seen one. It was in only ten metres of water and the huge brown monster eyed me and turned and swam into a cave. These fish were feared more than sharks by the early pearl divers, and I wasn't about to follow it home. Pearl divers may have had confrontations when they encroached on the grouper's territory, but spearfishermen usually had no such problems and the grouper had to be protected from them. The dive continued with the appearance of a maori wrasse, with saucer-shaped iridescent scales and a

A coral cave off Cairns on the Great Barrier Reef. Chris Deacon is hovering at the entrance to a very large cave with gorgonia growing in the low light levels of the cave.

Mangroves grow in estuaries where silt and mud has been brought down by rivers. They help to stabilise the land which is being deposited and enrich the bottom sediments. They provide a very productive and efficient environment being a home for breeding birds and marine animals.

A soft coral garden — its vibrant colours are a contrast to the usual pastel colours of hard corals.

Tropical waters are characterised by strikingly beautiful fish. The angel fish have dramatic colour changes between juveniles and adults which led people in the past to believe that they were different species. They are found singly and in pairs and can be hand fed. They approach quite close once they are accustomed to the divers.

Above: Sea snakes are curious creatures and these olive specimens have been captured with their tongues exposed in a rare shot. They enjoy looking at their reflection in a diver's facemask which frightens the wits out of unsuspecting divers. Although their venom is extremely poisonous, their short fangs would have difficulty penetrating a wetsuit, but one should not attempt to handle them — they could bite a carelessly exposed finger. The snakes are not encountered on most reefs but are common in certain southern reefs such as Swain Reefs, Keppel Island and the Yongala *wrecksite off Townsville.*

Below: While exploring Flinders Reef we observed a big school of fish feeding on algae covering the rocky reef in the shallows. As we approached, the school ignored the divers and continued eating without fear, making quite a mess in the water.

hump on its forehead. It is known as the Napoleon wrasse in the French Pacific colonies. It had apparently heard me and emerged out of the gloom. It was about 1.5 metres and 68 kilos and almost as big as me. It paraded back and forth before disappearing down the reef. I was looking at the gorgonia and black coral, when another big fish appeared. It was the spanish mackerel, a sleek silver-bodied fish, that came streaking into view, paused and almost as quickly streaked out of sight. In one dive I had encountered three spectacular reef fishes.

Butterfly Bay on Hook Island has many localities with coral bommies, caves and gullies and a proliferation of soft corals and batfish, maori wrasse and rock cods. Manta Ray Bay is another popular site on Hook Island where, in the winter, you are likely to see the mantas. At the Woodpile, a vertical wall drops from ten to thirty metres with undercut shallow caves and black coral.

Sea snakes live on the reefs in among the coral, particularly on the southern part around the Keppel Islands. Their venom is among the most poisonous of any snake but they have very small fangs that would not penetrate a wetsuit. They could, however, bite a misplaced finger, so care should be taken. They are just one of the many interesting creatures found around Great Keppel Island and the diving reefs such as Man and Wife Rocks.

Sea snakes, when first encountered, are quite a shock. Divers can be exploring the reef oblivious to what is going on in the distance. Dive leaders are usually the first to spot the sea snake, that is if the sea snake does not see the divers first. The sea snake will swim over to investigate the diver, who is suddenly confronted by a snake peering into his or her facemask and a minute examination of his legs and arms by the snake follows. It takes a very calm diver to stay relaxed and not shoot rapidly to the surface. When the snake disappears from view, the diver does not know whether it is crawling through his scuba gear or has gone, and must rely on his companion to tell him.

While the Great Barrier Reef extends only as far south as Lady Elliot Island, this is by no means the end of the coral or reef wildlife. Inshore of Fraser Island, live dugong, those fascinating mammals that gave rise to mermaid myths. Divers from Brisbane like to dive on Flinders Reef, off Moreton Island. The density of coral is not as great as on the Barrier Reef, but there are many species of fish which are common to both. This is also true of Lord Howe Island which is the most southerly coral island in the world.

Flinders Reef has its own personality too, with sights such as the rabbit fish that Kevin describes:

> I was swimming towards some rocky pinnacles that rose to the surface when a large school of yellow and silver fish surrounded the reef and began nibbling the algae growing on it. My dive buddy Warren Mahon swam into the school and they parted to let him through. Then a school of pike cruised past as I noticed some egg cowries and reached to pick one up. When I looked up they had gone.

Moreton Island is a favourite with Brisbane divers and one of the reasons is the artificial reef off Cowan Cowan, known as the Curtin Artificial Reef. The *Melbourne* was a thirty-five metre tugboat sunk on the reef in 1979. The wrecks have attracted corals, fish and animals seeking shelter or a surface to grow on. Large predators patrol the area feeding on the marine oasis. The surrounding seafloor consists of dunes of fine white sand, which are barren except for the life attracted to the wrecks. This area is reputedly the home of giant Queensland grouper, rarely seen on the wrecks since spearfishermen began to hunt in the region, but as Kevin describes, they are still there:

> When the tide reached full high and the current slowed, Rainer Pabst led the way down onto the wreck with Chris and I following. The *Melbourne* is in good condition, schools of coral fish darted amid the coral and along the companionways. Above the wreck a school of turrum relentlessly patrolled. We swam along the deck and paused at the stern. Looking at the sand five metres below, a huge black stingray swooped by. It rose like a spaceship and cruised out of sight leaving a vapour trail of sand. Chris lay in the depression it had left and her arms would not span its width. We peered into the gloom, the visibility was about ten metres. As if on cue, a huge shape materialised in front of us. Emerging from the edge of my vision, it swam straight towards us, then rested on the sand. It came closer, only two metres away, then turned and came back for a second look at us, visitors from another world. A second and third grouper appeared. Within minutes we were surrounded by seven fish, two metres in length and over 200 kilograms in weight. They were circling around us watching our every move.
>
> I moved out across the sand dunes to get better pictures, knowing the grouper would be less spooked by one diver alone. Out of the gloom another five fish were lying in a hollow between the dunes. As I approached, they rose off the sand and turned to inspect me. Beyond them was another sand hollow and another group. We had stumbled on the resting spot of a school of giant grouper, a phenomenon never reported before. We were on an underwater desert in twenty-five metres of water, surrounded by twenty giant grouper all larger than us, yet I felt no danger. I swam back to get Chris and Rainer and they too, were astonished. We stayed with the grouper until our air was low.

Brisbane dive boat Diversion I *passes the huge sandhills of Moreton Island.*

Divers on the Gold Coast witness sadly the demise of the big grouper from spots like Nine Mile Reef where they were once so common. These reefs south of the Tweed River are reached by launching a small boat at Rainbow Bay. Cook Island, which is just off the coast at

View of Mt Lidgbird from Mt Gower with Mt Eliza and Mt Malabar across the lagoon and the Admiralty Islands in the background. The slopes of Mt Gower are the home of the Lord Howe Island woodhen, which has been rescued from near extinction by a breeding program on Lord Howe Island.

Kingscliff, has an extensive reef with coral and small tropical fish, while further offshore Fido Reef has surgeon fish, red morwong, yellowtail kingfish, spanish mackerel and black cod. Five Mile Reef between Nine Mile and Cook Island is an extensive shallow area. While such offshore reefs are clearer, the current is often strong and then the inshore reef areas off Kingscliff, are favoured. A few hundred yards east of the bommie off Cook Island is an old wreck.

Further north along the Gold Coast off Palm Beach is a reef where the pelagic fish are the big attraction. In summer when Spanish mackerel are in the water, just after Christmas, dozens of boats will be seen on the reef. Divers can swim down to the level where the Spanish mackerel are swimming and hover there and the fish will circle, gradually coming in closer. This reef is protected from southerly winds and is about ten metres deep with little or no current.

Lord Howe Island is the southernmost coral reef in the Pacific, an island listed by the Australian Heritage Commission in the Register of the National Estate. The coral community on Lord Howe is in a delicate balance, attained by the periodic die-back as a result of cold, subantarctic currents, and the recolonisation by larvae brought down by tropical currents. There are thirteen species of fish that are found only in the Lord Howe area.

Lord Howe is the eroded remains of a large shield volcano. Some of the sands at the northern end of the island are wind blown sands and

Above: View of Lord Howe Island taken from Mt Eliza looking towards Mt Gower and Mt Lidgebird with the lagoon and North Passage visible as a gap in the reef. This is the entry point to the lagoon for yachts and vessels visiting Lord Howe. The lagoon is excellent for diving, particularly at Comet's Hole and Erscott's Hole which are past the island, visible in the far part of the lagoon, at the southern end. The outer reef area also provides very good diving along the reef face, an area not presently proposed for a marine reserve. There are a number of wrecks around the entrance to North Passage and the sand beach in the foreground is the site of the original settlement of Lord Howe Island.

Middle: A Lord Howe Island dive boat anchored at Admiralty Islands with Lord Howe Island visible in the background. A cleft in the rockface on Roach Island called 'The Eye' provides exciting diving.

Below: The deeper drop offs at the Admiralty Islands support a rich variety of gorgonia and soft corals. Divers can glide over the walls and drop offs enjoying a kaleidoscope of colour.

A special feature of Lord Howe Island are the coral canyons, arches and caves which result from the volcanic base on which the coral is growing. The sand filled gutters may belly out 10 m under the reef on the sand bottom. There may even be tunnels which the diver can explore. Sometimes these can be quite confusing, and may necessitate a trip to the surface to find out where you are.

were formed when the sea was at a lower level during the last Ice Age. The coral reef was more extensive previously than it is today and during times of lower sea level, was cut by water draining from the land which formed the reef passages and Comets Hole. Once a barrier reef with a lagoon surrounded the entire island but the reef and lagoon are now restricted to the western shore. The lagoon area has three outstanding dive sites accessible to divers, snorkellers and even glass-bottomed boats. The operators of these boats have been feeding sea urchins from the lagoon to the fish, with some concern by divers about the effects of indiscriminate harvesting of the sea urchins. The lagoon area is one section of a proposed marine reserve, to include Erscott's, Comet's and Sylph's holes. The other area which has been proposed is the deeper water to the north which would include Malabar, the Admiralty Islands and Ned's Beach.

Ned's Beach is a shallow water area where the fish will eat bread from your hand and whirl around you. Mullet, whiting, wrasse and silver drummers can be fed from the shoreline, or from the surface, or on the bottom. Extending underwater from the eroded cliffs of Mount Eliza are gradual slopes that extend into deep water. A deep cleft between Mount Gower and Lidgbird and Ball's Pyramid is approximately as deep as it is high, that is around 1000 metres. Ball's Pyramid is a rock pinnacle fifteen kilometres south of Lord Howe, standing on a shallow platform. The diving at Ball's Pyramid is similar to Lord Howe Island with abundant hard and soft corals.

A special feature of Lord Howe Island is the coral canyons, arches and caves which result from the volcanic base on which the coral is growing. The sand filled gutters may be ten metres high, bellying out under the reef on the sand bottom with tunnels extending in all directions. The labyrinth may be quite confusing sometimes, necessitating a trip to the surface to

The crimson soft corals which line the cave are a feature of Lord Howe Island.

find out where you are. Another characteristic of the diving at Lord Howe Island is the amazing variety of soft corals living on the reef. They are common in well lit, shallow waters and on boulders in contrast to the Great Barrier Reef where they are usually found only in caves. On the deep reefs of Lord Howe Island, black coral trees abound. The polyps are rusty in colour, contrasting with the soft pink corals.

Jeff Deacon is divemaster on Lord Howe Island and runs dive trips to sites in the Admiralty Islands, Mount Gower and the popular lagoon holes.

The east coast is not the only part of Australia with coral reefs. The west and north also have their share. In the north, diving around Darwin is rather restricted to getting to the offshore islands, off Coburg Peninsula. It is possible to obtain permission to visit Coburg Peninsula. Other sites are the offshore islands, near Darwin, which are accessible by charter boat. In the Territory, water clarity and weather is best after the wet season, May to June being ideal. Marlin and sailfish migrate from Bathurst Island north of Darwin across the Arafura Sea to the islands off Gove Peninsula and down the Gulf to Groote Eylandt, so the fishing potential of the Top End is great.

On the North West Shelf, an area currently being developed for natural gas and oil, there is enormous potential for a tropical fishery. Many young fish live on the Shelf fishing grounds rather than in coastal nursery areas such as mangroves. The Shelf is between tropical and subtropical influences and, in summer, cold nutrient rich waters wash onto it bringing in larval fish. Scampi or Norway lobster or langoustine, is a deep water crustacean species, which may form the basis of a new fishery in the region. The area has been fished in different periods by the Japanese and by the Taiwanese. The areas of Ashmore and Cartier Islands are under consideration as a marine reserve, but are presently being fished by the Taiwanese, as is Rowley Shoals, another proposed marine reserve. Since 1977 charter boats have been taking divers out to Rowley Shoals from Broome.

Rowley Shoals are shelf atolls isolated from each other by deep water. The reef rim encloses a lagoon with one or two passages. Mermaid Reef, the northern shoal, is sixteen kilometres long with the deepest lagoon of the three about twenty metres. There is no permanent land but a series of cays at low tide. On Clerke Reef, the middle shoal, an island known as Bedwell Island, projects above high tide level. It is unvegetated and is one of only two breeding sites for the red-tailed tropic bird in Western Australia. Imperieuse Reef is the most south westerly shoal and the largest. From April to October is the best time to visit the shoals. The reefs are comparable to the best coral reefs in the world with clear water and many species of fish not found elsewhere in Australia. Potato cod, coral trout, bêche-de-mer, trochus and giant clams are abundant in clear water. The sheer drops and sheltered coral lagoons rival the coral reefs of the Great Barrier Reef.

Thought to be the source of mermaid myths, the dugong is almost extinct in some areas, but in Western Australia it thrives in the protected waters of Shark Bay on the mid west coast.

Ningaloo is a coral reef system on the coast of Western Australia extending north from Point Murat round North West Cape to Amherest Point. Unlike the Great Barrier Reef, it is close to the coast yet it supports a wide variety of marine life, in close proximity to a very arid land area. Commercial development in the area is centred on Exmouth and Coral Bay. Commercial trawlers fish for prawns out of Exmouth, a town opened in 1967 to house families of men working on the US Naval Communications Station at North West Cape. Bundegi Reef is reached from the road out to the base. It's 100 to 200 metres out to the reef, which is rich in marine life including egg cowries. Off Lighthouse Bay is the

The clown fish snuggles into the tentacles of the anemone when he feels threatened, prodding the tentacles which may close around it for protection.

wreck of the *Mildura*. The area is a paradise to explore, if you like to be off the beaten track.

Steep Point is Australia's westernmost extremity, where the Indian Ocean swells break against a sheer cliff. It was here that the Dutch East Indian *Zuytdorp* was wrecked in 1712. Steep Point is the northern end of a cliff line that extends to Murchison River, 200 kilometres south. To the north across a strait is Dirk Hartog Island where the Dutch explorer, the first European on the Australian mainland, erected a plate. The plate now resides in the museum in Amsterdam. Between the Island and Steep Point is a winding natural channel to a multitude of rocky bays with sandhills behind them. South Passage forms a dog-leg channel leading into Denham Sound, one of the large expanses of water contained by the islands and peninsulas of Shark Bay. There are an estimated 2000 kilometres of coastline between Carnarvon and Steep Point, while the distance as the crow flies is only 153 kilometres. Denham is the main town, while Useless Loop is a privately owned town, the site of Shark Bay Salt Pty Ltd, who extract salt from the ocean in evaporation ponds. The diving in South Passage includes three different environments. The shallow headlands and rocky bays that line the passage offer good snorkel diving with large undercut ledges and small caves with a multitude of tropical fish, as well as southern fish like wrasse and snapper, while pink snapper and norwest snapper, or spangled emperor, are often found in mixed schools. Moray eel, butterfly cod and cuttlefish are found on almost every headland and on the sandy bottom, in only three metres of water, are large stingrays and shovelnose sharks, which congregate near the headlands where stronger currents flow past on the changing tides.

The bottom reefs of South Passage receive the ground swell that rolls relentlessly around Steep Point. Most of the reefs in the passage are on the north western side and extend offshore approximately one kilometre into the centre of the passage. The clearest water and most abundant marine life is found on the deepest reefs in the middle of the passage. Here limestones ridges extend to within two metres of the surface. There are very few caves but there are large ledges, with occasional brain or encrusting corals, but the big attraction is the fish. Jeff Mullins, an underwater photographer, describes a dive on the reef:

> On a recent dive I was greeted at the bottom by a pair of coral banded shrimps, peering inquisitively at the boat anchor which had hooked up in their rocky home. Not far from them was a large ledge, surrounding the lump of reef we had spotted from the surface, the whole surface of the reef appeared to be coated in small tropical fish, all going about their everyday business. Fish such as beaked coral fish, pullers, small wrasse, damsels, and butterfly fish of five or more varieties. Around the base of the reef, a small group of baldchin grouper picked amongst a few broken shells on the sandy bottom. As I approached the darker shaded overhang, I could see the large legs of several jumbo crayfish and they came out from their protective ledge to check me out. Something made me turn around at this point. I was being watched by a dugong only two metres behind me, peering over my shoulder. There was another just behind him. The closest dugong, circled me, then quickly disappeared. A few minutes later I was circled at very high speed, then they took off. It was a few minutes before I realised I had been buzzed by an animal which is rarely seen underwater.

The dugong is fairly common in Shark Bay where, unlike other areas of northern Australia, it is not hunted by the Aboriginal people. The dugong feeds on the seaweeds in the shallow waters. Its 'elephant man' face has evolved to tear up the seagrass on which it grazes, like a cow in a paddock.

A long line of white breakers extends one kilometre south from Surf Point on the south west of Dirk Hartog Island. Even on the calmest day the sea breaks along the reef line. The leeward side of the reef has the clearest water in South Passage about twenty metres. There are huge clumps of staghorn coral, large plate corals, and covering the sandy bottom are several varieties of soft coral. Bullseyes shelter behind ledges, crayfish peer from caves and black-tipped cod, breaksea cod and sand bass are found in the recesses. Out amongst the coral are the tropical-beaked coralfish, parrot fish, humbugs and the occasional clown fish and anemone. Dugongs and turtles congregate in the shallows at the back of the reef in the late afternoon, perhaps to avoid the large sharks that prowl the passage at night. Where the reef joins Dirk Hartog Island, there is a large sandy bay with patch reefs, where grey reef sharks congregate darting into the shallows, attracted by the schools of whiting.

Shells are abundant and include egg and tiger cowries found among the soft corals, and bailers, spiders and clams, on the reefs and sand. At

One of Western Australia's unique fish is this red scorpion cod, living on an artificial reef made of old car and truck tyres at Rockingham beach, 40 km south west of Perth.

The batfish is common in tropical waters and is often found in schools which approach divers while they are decompressing near the surface. They readily accept food from the divers and are also a popular aquarium fish.

Thousands of silver bullets burst in an explosion of silver from the diver as she explores the coral bommie of the reef face. Feeding frenzies of bigger fish scatter these schools.

Monkey Mia on the east side of Peron Peninsula twenty-seven kilometres from Denham in Shark Bay, the wild dolphins come in. They come from the surrounding waters every day to be hand fed on the beach front. They like to be patted on their sides and fed and are the delight of holiday makers.

The dolphins of Monkey Mia are wild animals, free to roam the seas unrestricted, yet they return daily to the same place. Some controversy has been expressed about the capture and training of dolphins and their restriction in oceanariums. The dolphins of Monkey Mia show that it is possible to see wild animals unrestricted by humans.

Cape Peron at the northern tip of Peron Peninsula has many shallow reefs while Coral Patches in the Denham Sound has many reef fish in limited visibility. Remote Cape Inscription on the northern tip of Dirk Hartog Island is the site of Dirk Hartog's visit and where he erected the post and pewter plate. The best time to visit Shark Bay is from June to October when windless conditions bring very clear water.

The Abrolhos Islands are located off the mid-west coast of Western Australia, south of Shark Bay and out from Geraldton. They extend for eighty kilometres to a point due west of Port Gregory. The islands are in three main groups — the Pelsaert Group is the most southerly, Easter Group in the middle and Wallaby Group in the north. North Island, to the north west of the Wallaby Group, is an adjunct to the others and distinct from the groupings. The islands are coral rubble deposited on existing coral reefs. There are professional fishermen's huts on most of the islands, a refuge for crayfishermen working the islands from March to June. The area is

Parrot fish will often be found under ledges on night dives. They cover their bodies with a cocoon at night which is thought to be a protection from predators such as the moray eel which hunts by smell and touch.

an A-class reserve, so permission to land on the islands must be obtained from the Fisheries and Wildlife Department. There is no fresh water nor any facilities except those available only to professional fishermen. Boats can be chartered to dive the Abrolhos Islands, or arrangements can sometimes be made with fishermen to use their facilities and boats to gain access. Once you have overcome the difficulties of getting there, the diving is spectacular.

On the east side of Pelsaert Island a drop off extends from six metres to twenty-five metres. The shell and coral rubble bottom extends to the drop off, then the wall is covered with hard corals and fish such as butterfly cod, long-snout boarfish, Westralian jewfish, and baldchin grouper, come up to you. The coral bommies at the base of the wall are home to crayfish and whaler sharks, zebra angels, goatfish, sergeant bakers, coral trout and mackerel, fish which may be sighted in water visibility that ranges from twenty-five to forty-five metres.

Half Moon Reef is a good-weather dive site only, the leeward side is big game country and anything from a shark to a grouper is a likely visitor. There are gullies cut into the sloping reef front and short kelp adorning the top of the ridges, with coral trout and jewfish in the caves and pelagic fish such as Spanish mackerel, shark mackerel, yellowtail kingfish and samson fish buzzing by with the sharks not far behind. Iris Refuge Island is tiny, but the reefs surrounding it can be dived in strong southerly or westerly wind conditions, as they are inside the huge lagoon formed by Pelsaert Island and Half Moon Reef. There are coral bommies rising from the bottom in thirty metres right to the surface, and the water is clear with no surge. Damsel fish, clown fish, parrot fish, staghorn and plate corals make this a good spot for the photographer. The charter vessel *Horizon* uses Iris Refuge Island for a base. The east side of the Easter Group has drop offs which are almost vertical down to thirty metres. In places there are holes in the reef where the coral and fish life is delicate and abundant. Rolland Passage on the eastern side also has even deeper drops offs. In Suomi Island Passage the *Nellie Melba* and *Horizon* have moorings that they use when visiting the island group.

In the Wallabi Group lies the *Batavia* wreck, on the southern end of Morning Reef. Jeff Mullins describes a recent dive on the site:

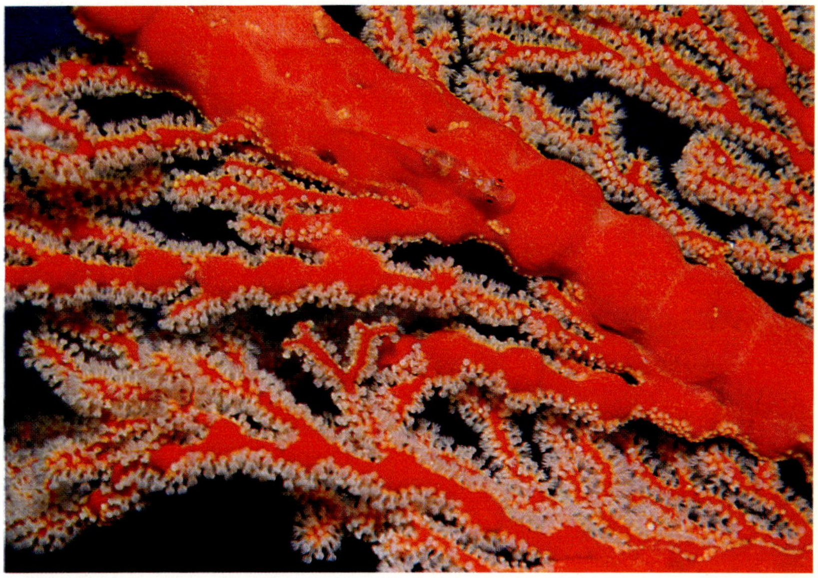

A minute fish resting on the gorgonia has adopted the colour of its habitat.

> We entered the water from the boat anchored 100 metres from the reef front and followed the chain laid along the bottom by the West Australian Museum into the wrecksite. A large hole in the limestone bottom about eight metres deep with one or two cannons on the sides was all that could be seen. Swimming further in, we came across two anchors lying in the shallower parts of the reef, about three metres deep, where the surge was so great we could go no further. What a let down after all I had read and seen about the *Batavia*. It appears the West Australian Museum has removed everything except what we had seen, for the purpose of rebuilding it.

A short distance away on the northern side of Morning Reef is a very good site for snorkellers and photographers. The reef has gutters and holes that drop to eight and nine metres, like giant aquariums, with very tame fish and clear conditions.

Australia has some fascinating places to interest divers from anywhere in the world. The many variations in the type of terrain and marine life, give endless possibilities for exploration, but it is the coral sea diving which attracts divers from all over the world. However other Australian diving areas are almost as famous as the Great Barrier Reef. In later chapters such sites as Piccaninnie Ponds and the Nullarbor Caves — well known, after many documentaries that have been distributed worldwide — will be explored.

In Jervis Bay the Gorgonia Wall is a favourite site for divers. Its sheer wall drops to 30 m. Looking up towards the surface, diver Andy Fraser can be seen hovering beside the wall.

SOUTHERN SOJOURN

Though at first glance the waters off southern Australia may appear devoid of many animals, the reefs and seagrass meadows actually provide a rich habitat for an amazing array of inhabitants. Indeed, the diversity is such that, of all the world's temperate seas, Australia's southern oceans contain the greatest number of different species.

Apart from coral reefs, the most outstanding feature of underwater Australia would have to be the expanses of giant kelp that border the eastern, southern and western coasts of the continent. Like a huge submerged forest, the kelp rises 30 metres from the sea bottom. Exploring this forest floor divers can expect to find the enigmatic weedy sea dragon, red and green striped senator fish and even the sedentary draughtboard shark hidden among the protective fronds of this nutrient-rich sanctuary.

On the narrow strip of seabed bordering southern Western Australia, South Australia and Victoria, divers will find savannah-like stretches of seagrass and seaweed. Animals and plants in this environment form the basis of the food chain in southern waters. The southern reefs are another interesting feature. Though not as colourful as their tropical counterparts, these reefs nevertheless harbour rich communities of fish and possess a charm and beauty all of their own. Red gorgonians, pink seawhips, purple anemones, grey soft corals and white sponges, as well as red fox wrasse, black, white and yellowbanded eastern morwong and incandescent blue devil fish, are just some of the colourful yet wonderfully diverse residents found on southern reefs.

"I was spearfishing off Beecroft Headland, Currarong, on the northern headland of Jervis Bay. We were on the bombora, a rocky reef, 200 metres off the headland. I dived to the bottom in fifteen metres of water to photograph my brother, Jeff Deacon, who had just speared a large kingfish. As I returned to the surface for air, I felt a tremendous grab on my leg. I looked down, thinking it was Jeff or Phil, and saw my right leg in the shark's mouth. I instinctively began to hit it across the head with my right fist and struck it in the eye. With the other hand I hit it with the bulky camera. The shark's reaction to this onslaught was to immediately release its grip. It had not had sufficient time to shake its head and vibrate its body, which would have cut through my leg in seconds. It swam in front and slightly above me and it seemed to take ages to pass. I released my lead belt and was very short of air, but the belt caught in my camera. I untangled it, because the camera was my only weapon if the shark returned, then swam to the surface.

I had no air to clear the snorkel and spat it out. I was watching below me to see if the shark was returning, but he was nowhere in sight. I swam over to Phil and explained I had been bitten by a shark, but he did not understand and went back to his fish on the gun. My brother, Jeff, when approached, swam down to cut the fish off his speargun cord and I thought he had not understood. I headed for the dive boat with Jeff close behind. I felt a great sense of relief when I got back into the boat. The gashes in my leg were of no consequence to me. When we got back to the launching ramp, a fisherman approached and asked what had happened. Not wanting to go into a long and detailed account, I told him I'd slipped and fallen on the rocks."

The backlit headland of Crocodile Head makes this afternoon photograph seem like a moonlit night. These steep cliffs plunge beneath the headland to depths of up to 40 m.

Late in winter Port Jackson sharks enter Jervis Bay in their hundreds, lying on the bottom in gutters and caves. They can be readily approached and handled as they are harmless, unlike the white pointer shark that attacked Kevin. Wherever the diver swims around the coastal fringes of the Bay they are found gathering for mating and egg laying. The eggs can be found in early spring in places where the Port Jackson sharks previously massed. Their corkscrew-shaped egg cases are often washed up on the beaches with the seasonal die off of seaweeds. Offshore winds bring calm seas during winter and it is then that divers explore the caves and caverns of the cliff faces of Point Perpendicular. These same westerly winds provide clear water as particles of sediment drop to the sea floor leaving calm, clear waters and pleasant diving.

During winter, large schools of fish congregate under the cliffs and drop offs of Point Perpendicular: red morwong, yellowtail, nannygai and many pelagic species. They gather in schools dispersing with the onset of summer. This is just one of the many seasonal changes at Jervis Bay. During early winter, the giant Australian cuttlefish enters the bay and the reefs offshore. Divers may see them mating and egg laying. Later on, in the winter, huge schools of kingfish, nannygai, pike and yellowtail assemble. The pelagics group together in layers from forty metres to the surface and as the divers descend through the schools of fish surrounding them, the schools move and shimmer in swirls of action. Later in the season, as the sea temperature increases, they disperse.

As summer approaches warm currents infiltrate the coastal waters bringing crystal clear water called the 'blue current'. The clearest water comes from both the blue current and the calm conditions of winter. Visibility increases to forty metres and schooling northern pelagic fish appear. Occasionally the huge fins of sunfish, looking like giant sharks on the surface, are sighted. This fish has been noticed often, in the Jervis Bay area where it comes close inshore on the warm currents. It is a gentle giant lazing on the water surface, its huge form is easily approached. Dolphins are found throughout the season but schools appear from time to time with large numbers of individuals migrating into and out of the area. Whales are sighted on their seasonal migration. A most sensational personal encounter with dolphins was at Currarong when Kevin and a group of dive school pupils were in a sandy bay in shallow water. He looked up and observed fifty or more dolphins coming in from the sea. The students discarded their scuba gear and snorkelled with the dolphins.

The Port Jackson sharks congregate around the foreshores of Jervis Bay in large numbers to breed during the winter months. They are not aggressive when breeding and can be handled. One or two hundred sharks could be seen on one dive. Their crushing plates are capable, however, of inflicting a wound if they are mistreated.

During spring many species of fish and shells can be seen laying eggs. Scallops are present in large numbers on the scallop beds of Jervis Bay and laws to protect them from overpredation have been implemented. One of our favourite Jervis Bay characters is Humpy who can be found at Summercloud Bay at any time of year.

Ray Gates is stroking the tentacle of the giant Australian cuttlefish as Paula Gates looks on. The cuttle is relaxed — as his colourations affirm — while the visiting American divers stroke him.

Bubble Cave, a phenomenon of divers' exhaust, is seen as it collects in the roof of the cave as an underwater mirror. Like 'Alice Through the Looking Glass', divers can pop their heads through the mirror, take off their masks and talk.

Above: The Docks is one of many dive sites at Jervis Bay which during summer, provides sheltered diving from north easterly winds as it faces south. The Docks' terrain offshore is made of reef sponge gardens and caves. It is a habitat for many marine animals and a popular site for both beginners and experienced divers alike.

Below: The diver feeds a blue grouper which has been a protected species for a decade now. It was speared in large numbers in the early days of diving when its habit of swimming up to divers made it easy prey.

He is a ham. To find him you have to dive to ten metres, about that distance offshore. Humpy is a southern blue wrasse, known as a grouper. They were popular with spearfishermen before they were protected. Humpy had a run-in with a spearfisherman and was hit by a spear in the back, which gave him the hump on his back. He has become a disruptive part of the dive school at Jervis Bay, as Kevin describes:

> We were settled on the sand practising regulator removal in six metres of water, the divers in a circle facing me. I noticed Humpy had joined the group. He swam up to the class, settled on the sand and watched. Then he went up to each diver, one by one, completing a full circle around the group, and ending up back where he started. When he got back he sat patiently for a while, then decided it was time he was fed. He started swimming back and forth amongst the dive class, disrupting the exercises.

The discipline of the class was destroyed. Chris tells how he is pacified:

> We've thought about feeding him raw fish, but sea urchins are handy and are what we've always fed him. He recognises the urchins. As soon as you

pick up an urchin, Hump's there. He'll give you a nudge if you're not getting your act together quickly enough. He'll only stick around as long as the food's on, unless he's in a good mood. He can be overfed and then he's not interested any more. There's always a purpose to his visit. He can play around for a while, but when he's really hungry, he'll charge in and really get into the action. Everything in the sea is slimy, particularly the fish, but he's nice.

Recently Humpy has been joined by another blue wrasse, he is obviously male as the females are brown, and he comes in to feed too. His territory is different from Humpy's and when he infringes on Humpy's territory, Humpy attacks, chasing him out of the area.

Ulladulla, further south has sponge gardens and reefs similar to the Jervis Bay region but lacking the extent or range of diving. Sullivan's Reef, which is also known as Home Bommie, is located in an easterly direction straight off Ulladulla Harbour. It is a rocky reef that starts three metres down and drops to a sandy bottom in thirty metres. An extension of this reef to the north is called Golfcourse Reef, but it is totally different and separated by sand. There is quite a lot of fish life but it does not have the sponge beds and gorgonia that are a feature of Sullivan's Reef. The rocky Home Reef is dissected by caves and swim-throughs. A blue wrasse or grouper has lived in one of the swim-through caves for fifteen years now. He swims up to meet the divers as they come down the anchor line. He'll nudge them to be fed and will follow divers, going from one group to another waiting. Marine life is very prolific especially on the bottom of the reef. The top of the reef is kelp and drop offs extend in steps from eight to fifteen metres, then fifteen to twenty-five and so on. The floor of broken rocks and boulders is covered in sponges and gorgonia growing in the open. The fan shaped clusters come from a single base and cover whole rocks. They are easily damaged by boats' anchors, so divers should anchor elsewhere. There are lots of fish, yellowtail, mado, sweep, roughies and different wrasse, the brown and blue grouper, which are male and female.

Lighthouse Reef at Ulladulla is subject to seasonal changes. In winter it is fairly barren but in the summer there are schools of fish, and during autumn cuttlefish come in to mate. An extension of the Reef in deeper water thirty to forty metres is sandy. Abundant with sea whips and pelagic fish, it is covered in vase sponges, fan sponges and finger sponges.

The giant Australian cuttlefish is raising his tentacles in warning as diver Colleen Lander approaches.

Broulee Rocks, near the entrance to Burrill Lakes, is quite an exciting dive. It is the extension of the headland into the water about 300 metres in an east west direction and separated by sand. It is a bomora that comes to the surface and breaks in bad weather, and a flat plateau extending quite a distance in a westerly direction with sheer drops off the sides. There are many barren spots but the best diving is found on the southern side with sponges and corals. The other side of the reef is not as good for diving except in

Seal Rocks, a fishing village on the north coast of New South Wales is also a favourite site with divers.

one spot, called the Hole or the Horseshoe, which drops in a horseshoe shaped hole sheer to thirty metres. On the north eastern side it drops from the weedy reef to ten metres where sponges and corals occur but no gorgonia, which is found mainly on the western side of the reef. Mados, small school fish and pelagics with wrasse, are abundant in the kelp but it is the weedy sea dragons and kelp fish that are the most popular.

Brush Island is another coastal spot on the south coast of New South Wales that is popular with divers because of its clear water. It is some distance from the outflow of any major rivers and has clear water as a result. The best diving is off the front of Brush Island where a sheer drop on the north east corner extends from the surface down to thirty metres onto flat bottom. Strangely, kelp is growing to the bottom much deeper than elsewhere perhaps as a result of the very clear water.

Wrecks abound on the south coast from aptly named Wreck Bay, off Sussex Inlet, to Twofold Bay, home of the whalers, and nearby Green Cape. An interesting wreck at Merimbula is the *Empire Gladstone*.

This ship struck Haystack Rock, south of the small coastal town of Merimbula in September 1950. It was headed for what the captain believed was a lighthouse, but was in fact a light on Merimbula wharf. When he sighted the lights of Merimbula and realised the ship was too close to shore, he swung out to sea, but the ship stopped with a grinding roar — the stern was caught on a submerged reef. The crew remained ready to abandon ship all night, the people of Merimbula being unaware of their plight.

The *Empire Gladstone* immediately became a great tourist attraction. Four thousand visitors a day were reported at the site. While many of the town's fishing boats were transporting visitors to the wreck, others were busy removing 159 car bodies from the deck.

Today when you jump into the water over the wreck, the propeller shaft leads the way to a square cabin area, from which peer many species of fish that abound on the wreck. On the sides of this cabin are large holes, as I once found:

> As my eyes became accustomed to the gloom, I realised the seafloor inside the cabin was partly lined with large cobblestones. Holes in the roof made dappled shadows on the sand and at the far end of the cabin was a companionway. Beneath it I discovered a prize, a very large spanner. My air bubbles collecting on the surface formed a mirror reflecting the eerie scene. As I swam out of the cabin, my bubbles were escaping in a silver stream from holes in the cabin roof. The outline of the hull was still discernible though sand and rock filled many parts of it. The prop shaft guided me to a maze of conrods and engine parts.

Seal Rocks on the New South Wales north coast has plenty of fish life but sponge gardens and corals are rare. Shark Gutter, out on the rocks, is a big gutter filled with grey nurse sharks resting on the bottom and moving backwards and forwards. The Gutter is sheer, dropping from three metres into twenty-five metres with weed near the surface and rocks on the bottom. Schools of yellowtail swarm among the sharks, and a slight current brushes past. Another north coast diving area is Forster where Latitude Rock has a grey nurse gutter, and occasional tiger sharks are sighted. Where the current is stronger, sponge gardens and more prolific life is found on the offshore reefs. Grey nurse sharks are unique creatures that inhabit the southern waters in Australia. Divers and naturalists have dispelled the fearsome reputation that these sharks had. We now know they prefer a diet of kingfish. They live in gutters where fish are plentiful and where, because of the shape of the gutters, water is directed past them. Thus they are able to lie on the bottom with water passing over their gills without having to swim continually to obtain the same effect. Large numbers of these sharks were destroyed by early spearfishermen who

The dive boat anchors under the steep plunging cliffs of Jervis Bay's northern headland.

The Solitary Islands near Coffs Harbour offer subtropical diving in a remote area. South Solitary Island, the only one inhabited, has a lighthouse and a keeper's cottage.

Wobbegong, Port Jackson and grey nurse sharks are common inhabitants of southern waters.

thought they were destroying an enemy in the sea. Seal Rocks is one of the few places where they can still be seen. The gutters and caves, out from Seal Rocks, are now a popular dive location with never an incident of aggression from the inhabitants. The sharks have now been placed on the protected species list — a unique situation in Australia where they had been so feared. An animal once considered our greatest enemy has had to be protected from man himself: a sad indictment of the conservation status on earth.

The first encounter with a grey nurse is an adrenalin boost for divers. Even when they know it is not a maneater, it has the appearance of one. Its long, sloping mouth and full set of teeth does not create confidence for the inexperienced but its hydrodynamic shape is fascinating.

Coffs Harbour is the end of the warm tropical current on the coast. The North Solitary Islands are much warmer than the South Solitary Islands. A definite natural barrier occurs between the two Island groups. This can be felt in the water temperature and is reflected in the differing marine life in the two areas. North Solitary is the most exciting diving. Shark Bay, on the northern side of the Island, drops from ten metres to thirty metres with everything in the bay including grey nurse sharks, anemones and clown fish covering the bottom reefs. Huge kingfish surround you as you enter the water. Mushroom corals, brain corals and tropical fish abound. One exciting dive explores the tunnel in Shark Alley which goes into the middle of the Island opening into a big cavern. The bottom is sandy and not much marine life is seen until you come upon a shark buzzing out in the opposite direction: a thrill for the uninitiated. It becomes darker as you go in, and a torch is needed. On the western side is Anemone Bay where, from thirty metres to the surface, the rocky reef is covered with anemones and clown fish, all different types. They are so closely covering the entire floor of the bay that you cannot put a fin on the bottom or drop an anchor without landing on them. An occasional grey nurse shark can be seen, but as you come around to the southern end of the bay, the anemones stop abruptly. The southern end of the Island has two big gutters that drop sheer from the surface to the bottom.

The eastern side of North Solitary drops from the surface to over thirty metres only a boat length from the coast. Skull Rock to the north is very good diving too, accessible from Red Head over the bar, or from Wooli. From Coffs Harbour, the North Solitary Islands are accessible in good weather and the trip takes about an hour.

Around South Solitary, which has a lighthouse and cottages, the diving is deeper and off the Bantree on the south west of the Island, debris from the cranes that used to lift the stores to the lighthouse lies on the bottom. Anemones are fairly prolific and the bottom drops in boulders to thirty metres. Further around to the west is a bommie with a strong southerly current. This area is prolific with tropical fish orienting towards the current which runs at two knots. On the north eastern side is an exciting spot called The Arch. If the weather is calm, divers can swim through The Arch keeping it on their right then proceed the thirty metres to the other side of the headland. On the northern exit point,

The basking shark is seen with gills flared as he filters the plankton through his mouth.

grouper and cod abound. A big brown cod followed us up the anchor rope here. Sharks are also found around the Solitary Islands including hammerheads, tigers and whalers, and dolphins are also prolific.

South Solitary is closest to Coffs Harbour and visibility is not as good as it is around the northern Islands. Kelp is abundant but there are a few areas where the kelp is scarce and the diving is better with more fish life. Black coral is found in the gutters that look like angular gouges in the rock fanning out onto the sand. The northern side is quite barren compared to the southern side.

At Port Macquarie the cod hole lies just offshore and there are many sponges and small fish on outer reefs dived by local divers. A short drive away are South West Rocks where Fish Rock Cave is popular. It is a short distance by boat from Trial Bay. The reef comes out of deep water in excess of thirty metres and off the wall in the north east, depths of sixty metres or more are found. The dive through Fish Rock Cave is from the southern side of the rock. Divers drop into thirty metres of water where they enter a large split after descending down a cliff. The tunnel penetrates the entire length of the rock, gradually getting shallower as you ascend a chimney following a rope that has been installed to aid divers. Crayfish peer from crevices surrounded by walls of bright yellow sea anemones. The diver comes up the chimney into the shallow area where it ends in an air chamber, and companion divers can have a chat and exchange impressions of the dive whilst still underwater, in an air bubble. The sunshine anemones make a beautiful yellow orange display as you enter the air pocket. The exit from Fish Rock Cave is surrounded by the anemones and white gorgonia corals filtering pale blue light streaming into the cathedral-like cave. Looking out, huge schools of fish and wobbegong sharks can be seen, and the re-emergence into the light gives a sensation of rebirth.

Fish Rock is one of the South West Rocks. Another popular dive is the Wall, which is part of Fish Rock. This plunge into sixty metres brings you close to large cod and pelagic fish. Further north is a spot called the Shark Gutters where grey nurse sharks can be seen on days when the visibility is good from the surface.

The velvet weedfish is another dweller of the seaweed and camouflages well into the weed on Portsea Pier where it was found in 2 m of water.

The southern blue devil fish is found throughout southern Australia. In Port Phillip Bay it is found in the caves and ledges of the rip at depths of 10 to 30 m. Juveniles spend most of the time in the recesses of dark caves. This one is easily recognised as a juvenile, by the handsome black band around the tail and tip of the second dorsal fin.

Port Stephens is a beautiful region within easy reach of Newcastle and with some spectacular scenery just recently incorporated into a marine reserve. The Fly Point Halifax Park Marine Reserve is on the southern side of Port Stephens in a very accessible area for divers. Across the seagrass beds, the water drops into ten metres in a series of ledges covered with a variety of sponges and sponge encrusted spider crabs. Large bream will feed from your hand, if you take along some food for them, and large blue grouper wrasse are curious about divers. The area is dived on slack tide, preferably high tide, when water visibility may be up to twenty metres. Only a short distance away is Halifax Park where there is even a cleared path through the small boulders on the beach to reach the water. A rocky bottom shelves out to six metres then drops to fifteen metres and flattens out at twenty-five metres. Over the bommies covered with *Tubastrea* coral are hovering bullseyes and stripies. Among the sponges on the drop off are brittle stars, crabs, shrimps and octopus, and an occasional Moorish idol will drift by.

Byron Bay has offshore diving at Julian Rocks where black coral trees are found and there may be exciting encounters with all sorts of marine life. Julian Rocks are a series of rocky reefs extending out from the lighthouse at Byron Bay, the most easterly point of Australia. The rocks are a habitat of tropical and subtropical fish. The great variety of species mix in the area with black coral, anemones and butterfly cod. The black coral trees are on the western side facing the mainland. A family of large cod are found in the cod hole.Large bat rays and grey nurse and wobbegong sharks are local inhabitants. A number of reefs near Julian Rocks in the Bay provide good diving for visitors.

Butterfly cod is widely distributed in Australia. It is called by a variety of names, zebra fish, fireworks fish, lion fish, scorpion fish and dragon fish — names that indicate its venomous spines and the effects on those who touch them. It feeds on crabs, shrimp and small fish, as it drifts along the reef almost like an old sailing ship with majestic sails outspread and pennants flying.

Jeff Deacon was helping Kevin photograph a butterfly cod at Byron Bay when he was stung. One of the pectoral fins pierced his thumb as he manoeuvred the fish, it lashed out sidewards and escaped. Jeff got into the boat and applied a tourniquet which inhibited the flow of blood. The pain was bearable. When they got back ashore, they removed the tourniquet and within seconds Jeff collapsed. By the time they reached hospital Jeff was in considerable pain. In the early days of diving, stingings were not often acknowledged, so specific treatment was not readily available. Today, with increased understanding of the nature of protein-based toxins which respond to warmth, treatment is quick and effective. He was treated for an unknown toxin or snake bite. At the time, he described his hand as feeling as if it were stuck in a furnace but by the next day the swelling and pain had subsided.

On the south coast of New South Wales, Bermagui is the jumping-off point for Montague Island which has large shallow reef sections and deep drop offs but its most famous feature is the Seal Colony, as Kevin describes:

> When the dive boat arrives, the seals leap into the water. Once in the water they hang beneath the surface popping their heads up to watch the divers in the boat. As the divers enter the water, the seals swim up and greet them in a whirlpool of bubbles, fins and antics.It is a glorious experience,interacting with the seals. The large male bull seals guard the territory closer to the island and these huge animals will charge at the diver who rarely calls their bluff. As they weigh 500 pounds or more each, and charge right to the face plate, one is inclined to respect their territory. Out in the deeper water, one can frolic with the females and pups without danger.

Further south is Eden and Twofold Bay and Green Cape, a jutting point with diving in calm conditions possible on one side or the other. There are a large number of wrecks in the area. The *Ly-ee-moon*, the *City of Sydney* and the *New Guinea* are just some of them. The *Ly-ee-Moon* was a paddle wheeler that went ashore with seventy-one lives lost in 1886. The waters around Green Cape offer clear conditions and interesting diving.

In Sydney, where waters may be devoid of some species by day, there is plenty to see at night. The colours which may seem muted in dirty water, glow in a spectacular fashion under the diver's lights and the limited visibility is not so restrictive when the visibility is only as far as your torch light extends. Other divers armed with lights are easy to see and, as an extra precaution, in case a diver's light goes out, chemical glowsticks are attached to the diver's cylinders, visible to his companions. During a

The butterfly cod is known by various names, including lion fish, volitans, zebra fish, fire fish and lion fish — names which reflect its poisonous nature when touched. There are several species all exhibiting beautiful dorsal and pectoral fins supported by long thin spines. There are 13 spines along the back which is unlucky for anyone who handles them.

An inquisitive seal pup swims up to see what the diver is doing and is caught by the camera in a spotlight of the sun.

night dive, crayfish and other crustaceans are out walking on the reef, coral polyps are extended feeding, and basket stars have their arms extended reaching the abundant plankton. Fish asleep in caves and crevices can be approached and handled. Turning off the torch and watching your companion swim can be rewarded by a light show as the diver propels himself through the phosphorescence, or watching the cold light produced by tiny crustaceans. A school of dolphins, their bodies outlined in phosphorescence, streak through the water like dolphin ghosts with bursts of light.

Sydney Harbour has another unusual night-time surprise. On one dive several stony corals glowed in the light of an ultra violet lamp as we switched off our torches. A plate-shaped coral of uninteresting brown colour by day glowed by night with an eerie blue light, another was spotted with blue, while yet another glowed in bright apple green, and an alga glowed red as its fluorescence was unexpectedly revealed.

The marvel of fluorescence is the ability to change the wavelength of invisible ultra violet rays to the visible spectrum. Fluorescence has been traced to pigments in the coral tissue. Microscopic algae or zooxanthellae in the coral tissue, may use the light produced by the coral. Sometimes the light is produced by bacteria which may be present in many animals such as crustacea or even fish. Many other marine creatures have this property of fluorescence.

An entirely different, but equally fascinating, experience is provided by a little fish called the pineapple fish, as I found on a harbour night dive:

> Deep inside a cave a moving light is seen, only to disappear almost immediately. Our torch reveals a yellow pineapple, a fish with hard yellow scales looking for all the world like a pineapple. By switching off the torch, we see the light again, only to disappear before our eyes. It is the fish, we discover, as he opens and closes his mouth. On the lower jaw is a patch of tissue which glows in the dark. As the fish closes his mouth, the light goes out.
>
> The bioluminescence is like the light produced by the phosphorescence in the water. Bacteria in the tissue emit light as a by-product of their body functions. What benefit the fish attains by the relationship with the bacteria is uncertain. Perhaps little fish are attracted to the light as we were, and just maybe this slow swimming pineapple fish is able to catch them.

Divers too, use a few tricks to catch food.

Bare Island at the entrance to Botany Bay, Sydney, has spectacular and colourful life including this magnificent red anemone captured by A. Fraser on film.

Mussels, oysters and abalone make a delicious appetiser, but not many divers have tried the orange meat of the sea urchins they are so fond of feeding to the fishes. Yet the sea urchin roe is quite delicious and is a delicacy in Japan. Crayfish are the Australian equivalent of the Northern Hemisphere lobster without the claws. Marine crayfish come into the shallow waters around the coast to breed and can be found hovering in the dark recesses of caves, or among the kelp beds in southern Australia, their feelers just visible. Tropical crayfish hide among coral, but the southern waters are where the most and the largest crays are found. Divers catch the crays by reaching into their holes wearing gloves, taking a firm grip on the cray and tugging it out.

Diving the southern Australian coastline is as unique and interesting as the north. Abalone divers became commercial divers overnight in the 1960s when markets in Asia expanded for this delicacy. They explored much of the southern coastline from Eden in New South Wales, through Victoria and Tasmania. Abalone are shellfish found under rock ledges, off which they can be prised with the help of a stout knife. The underside of the ear-shaped shell is the fleshy foot that does not look very palatable. The animal is removed from the shell by inserting the knife next to the shell and cutting the muscle

The southern rock lobster is a highly prized seafood and an important commercial species. If such large specimens are to continue to be seen by divers, strict conservation measures will be necessary. In Victoria, divers with an inland amateur fishing licence are permitted to take up to 4 crayfish per person per day.

An ascidian, found on the sides and under ledges of the south coast where it is an attractive sight adorning the rock face.

A view from Wilson's Promontory looking towards the Anser Islands. This national park is a favourite with visitors and divers who like to explore its offshore reefs and islands.

where it is attached. It is then gutted and the black muscle rim cut off. The remaining disc of white meat is sliced then pounded with a meat tenderiser until soft, which is the secret of preparing abalone. Cooking takes about half a minute in hot butter then it can be garnished with shallots. Abalone soup can be made over a hot campfire if you have plenty of time, for it takes several hours to extract the flavour from the abalone turning it into a white stock that can be added to a vegetable soup and is an excellent appetiser for a seafood meal.

One of the best areas around the south coast is Wilson's Promontory. It is a bit difficult to reach the best spots offshore. The Anser Group are nine kilometres from the boat-launching ramp at Tidal River and the sea is often rough. There are three main islands and three smaller ones. Skull Rock or Cleft Island is one of the larger ones and its awesome mass of granite, with sheer sides up to 111 metres and skull like shape, is impressive. There are huge caves on the north face that make the skull appearance. There are some underwater caves below the skull on the north side and seals use the rocks inside. On the east side of Skull Rock between it and the big island of Kanowna, the water drops to between thirty and forty-five metres with a strong current and very clear water. The bottom is strewn with huge boulders up to twenty-four metres high and covered with sponges, soft corals, ascidians, and lace corals of great variety for the underwater photographer. Anser Island rises to a peak of 152 metres with a bay on the south west where the granite slabs have broken from the cliffs and plunged to the seafloor creating a series of big caves. The marine life is extraordinarily rich with some sponges unique to the area. There is also an unnamed wreck in sand at twenty-one metres. When a southerly wind is blowing, the opposite point of the island is protected by a headland with big boulders falling down onto sand at thirty metres. Kanowna Island has been chosen by a colony of fur seals which have selected the two bays facing east with gently sloping rocks into the water where the seals can land, protected from the worst weather. At the height of the season there are thousands of seals on the island. The young pup seals are born with a distinctive black coat that has no underfur and is not as insulating. The pups will jump into the water on a hot day to cool off and the bay becomes a mass of black. On the Island, some of the adults will climb to the top of the highest point in the height of summer for some unknown reason.

Close to Melbourne at Port Phillip Island a very big colony of seals lives, and stragglers provide excitement on the buoys of Port Phillip Bay. The Bay, on Melbourne's doorstep, is swept by strong currents so most spots can only be dived at slack water. Slack water periods occur every six hours and in between, the current ranges from two to six knots depending how close you are to the rip. Slack water does not coincide with high and low water but depends on prevailing weather conditions. North east to north west winds tend to delay slack water and south east to south west winds mean slack water will be earlier. High slack water is the best time to dive because the Bay is filled with clear incoming water rather than churned up harbour dregs. The Rip, at high water, can have visibility from eighteen to twenty-five metres for diving. Because timing is critical, boatloads of divers can be seen at all popular spots waiting for the tide, all geared up and ready to go.

Pope's Eye is a semi-circular rock reef that was begun as a fort to defend the entrance to Port Phillip. Inside the water is around two metres and outside it drops to ten to eleven metres, making it an ideal spot for novices. It is part of the Harold Holt Marine Reserve, a

Seals mill around the diver as one delicately takes food from his hand.

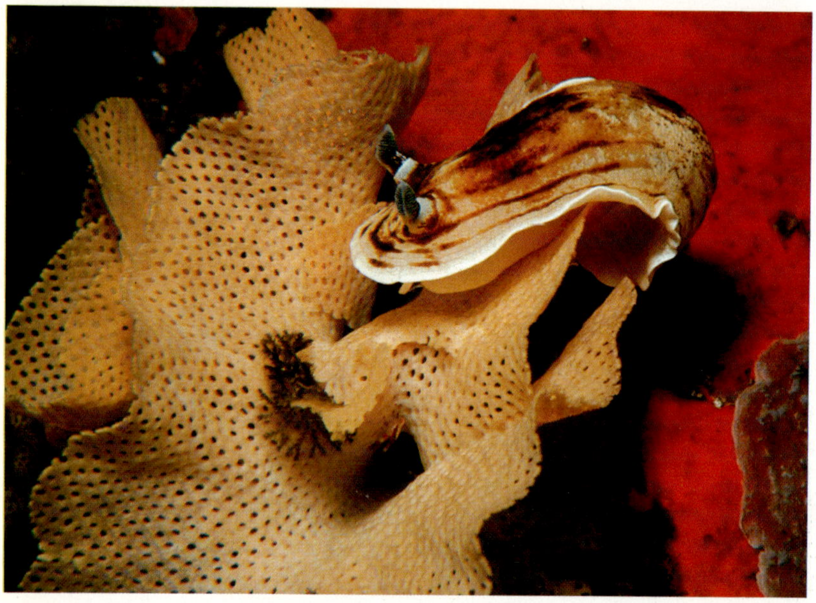

Above: A fish eye view of the world above, and below, Portsea Pier on Melbourne's doorstep.

Below: These common nudibranches are not as attractive as some of the more colourful nudibranches, but seen here in the lace coral, it makes an attractive composition.

scattered marine park encompassing sections of Port Phillip Heads and reefs inside the Bay. South Channel Island is another fort, which is primarily a snorkelling location, that has some spectacular gorgonian fans in only one to two metres of water hidden in kelp.

Diving the piers is a popular activity with Portsea Pier a favourite. Sea horses, pipefish and sea dragons cling to bryozoa and algae — the piers are havens of fish life and are interesting both by day and at night. Portsea Hole, which is northwest of Portsea Pier, can be spotted by the number of dive boats clustered over it on the weekend, all waiting for slack water. The exposed reef wall drops from thirteen metres to around twenty-eight metres and is dotted with ledges and huge caves. The fish are fed by divers — rosy wrasse, leatherjackets and green parrot fish are quite tame and may try to nibble the fingers of divers not wearing gloves. Blue devils, goatfish and colourful anemones make this an interesting dive in an amphitheatre hole.

The Kelp Patch between Queenscliff lighthouse and Point Lonsdale on the western side of the Bay can be dived as a current run. There are ledges in seven to ten metres with kelp fronds rising to the surface and the diver whizzes past bent, almost rendered horizontal by the force of the current — watching butterfly fish and herring cale dart in and out of the shimmering shadows. Another current dive is on the scallop beds in the Bay. Dive boats drop the divers in groups and the skipper watches for them to surface, then cruises up to them with a 'mermaid catcher' strung behind the boat on a line. The divers can haul themselves along the line or just rest from the current while the boat picks up other divers. Lonsdale Wall can be a more dangerous area depending on wind and tides, but is nonetheless an interesting dive dropping in a sheer wall from twelve metres to undiveable depths in a progression of ledges and undercut caves. The Wall starts at The Rip and extends into the Bay for nearly a kilometre. It marks the old path of the River Yarra. On the western edge it is flanked by Lonsdale Platform, a kelp covered reef in thirteen metres of water. At the twenty to twenty-five metre area are long-snout boarfish, magpie perch and butterfly perch in the caves. Slack water here occurs about three hours after high and low tide. Drift diving from a line attached to a buoy or to a boat is popular at times other than slack water.

Off Port Phillip Heads, in twenty-five to forty metres of water lies the ships' graveyard. In the 1920s and 1930s, a number of vessels were scuttled here including four submarines. A typical dive involves a quick trip to the bottom, a swim along the length of the sub, inspecting the rudder and stern and the torpedo tubes, then a leisurely ascent with a cautionary decompression stop. Penetration into one of the submarines is possible but any groundswell causes a violent surge at the open bow of the sub. The *Eliza Ramsden* which sank in 1875 at Corsair Rock at the entrance to the Bay provides another popular dive spot. The wreck lies on the southern boundary of the south shipping channel and it is not permissible to anchor on the wreck, so charter operators use a buoyed shotline to mark the wreck and let the boat drift during the dive.

Table Rock, Beaumaris, only twenty kilometres from Melbourne, is a safe dive for beginners. Here you will find plenty of marine life for macro photography: stony corals and soft corals, tube worms and feather stars.

A feature of the diving in southern Australian waters is the giant kelp beds that rise from the sea floor where they are anchored to the rock by holdfasts, to the surface twenty metres above. Living in the twilight world, beneath the spreading fronds of kelp that move on the surface in the swell, are tiny fish, sea urchins, anemones and crayfish. Stargazers, red velvet fish and handfish live in the kelp while the basket and plate sponges below the kelp are favourite haunts of the red rock cod and gurnard perch. Red biscuit starfish and other colourful sea stars, and nudibranchs, make diving the submarine forests of Tasmania a fascinating experience. Bruny Island on Hobart's doorstep, provides a natural barrier protecting the waters of D'Entrecasteaux Channel. The channel is fifty-six kilometres in length and from two to ten kilometres wide. It is bordered on the north by the Derwent estuary and Storm Bay, and on the south by the Actaeon Islands and is a good all weather area for diving near Hobart. Tinderbox is only a twenty-minute drive from Hobart and lies at the northern end of the channel. The strong tidal flow attracts fish such as barber perch, silver dory, and velvet and brown leatherjackets, butterfly perch, and rosy and senator wrasse. Dover, on the southern channel, as well as being the home port of many vessels of Tasmania's fishing fleet, is also the site of the wreck of the *Katherine Shearer* a barque that exploded in 1855 when nine tonnes of gunpowder was ignited. The proposal to make Ninepin Point in D'Entrecasteaux Channel a marine reserve led net fishermen to set up gill nets on the small reef. While politicians have vacillated, the reef has lost the larger fish that once milled around the diver in schools.

The algae that forms the undersea kelp forests of southern Australia is one of the sea's fastest growing plants, adding up to a metre in length to its fronds each day.

Bicheno is a town north of Hobart begun when whalers and sealers pulled into the bay for supplies. From about 1803 a vigil was kept on 'Whalers Lookout' for the pods of migrating

The kelp forests of Tasmania are much larger than their counterparts on the mainland of Australia. The overall length of the stems plus the fronds on the surface may reach 60 metres — a giant seaweed.

Beyond the reach of light which supports the kelp forests of Tasmania, are the deep reefs and submarine gardens. While they appear dull browns and blues to the diver, their true colours are revealed by torchlight in fantastic shapes and sizes.

whales. The Lookout is a granite knoll overlooking the township of Bicheno and The Gulch on one side, and the Tasman Sea on the other. The Gulch is the home port of a fleet of fishing vessels that supply the two seafood processing plants in the town. A marine research establishment is investigating the release of young abalone, crayfish and scallops which will increase the harvest from the sea.

Giant kelp forests predominate along the shores of Bicheno — a Jack and the Beanstalk climb to the surface from their holdfasts on the rocky reefs in thirty metres of water. Beneath the surface canopy of fronds, lies a shaded world of many creatures found usually at night. Schools of bullseye and trumpeter drift sleepily through the forest. Crested weedfish dart in to investigate, when a diver fins through the weed beds. At other times they are camouflaged in the weed, their bodies flattened against the fronds. The elongated fins even have holes to disguise the outline. These fish give birth to live young and when they are sluggish and fat may be pregnant. The velvet weedfish is another kelp dweller specific to this unique environment.

The brown algae that form these undersea forests are giant kelp, the fastest growing plants in nature, adding one metre to the fronds each day. Once lawnmower-like vessels equipped with cutters harvested the kelp on the surface, for production of alginate. The gel is used in dentistry, and food industries, even in printing industries where it is a thickening agent. Like many sea harvests, this industry has declined and the kelp grows dense on the surface.

Beneath the world of the kelp, in areas where light is too scarce for the kelp to grow, are sponge gardens. Although in the poor light they appear dull browns and blues, a torch reveals the brilliance of their colours. Sponges and sea whips are filter feeders which grow where a current sweeps past, but here they grow in still water. Living between twenty and seventy metres, the southern sea whip sends a flexible spiral off the reef into open water where its polyps can feed. Usually small jewel anemones live on rocky walls, but here they suspend themselves on the sea whips, removed from their competitors on the seafloor. Blue devil fish hide in caves during the summer, emerging in winter as adults, when they may be seen below the kelp forest.

The basking shark, a harmless plankton feeder, is just one of the many giants of these waters growing to a length of ten metres and weighing over three tonnes. The sight of a shark with a one and a half metre white mouth, coming towards you, is a bit disconcerting, but its ungainly swimming motion soon makes you overcome any fear. The most noticeable feature is the enormous gill arches that it flares as it takes in the minute plankton in gulps of water.

Port Davey in the most inaccessible part of Tasmania, the south west, is a drowned river valley. The Tasmanian government has just completed a survey of this remote area and declared it a marine reserve. Breaksea Island across the entrance to Bathurst Channel is darkly stained by the tannin from the button grass plains — a unique feature of the plant life of the south west of Tasmania. The light shafting through the water gives it an orange brown hue. The sea whips, gorgonians, corals, sponges and bryozoa, cover every millimetre of the bottom, so that it is impossible to avoid crushing them with flippers. The tannins in the water transmit the red end of the light spectrum, absorbing blue wavelengths. As water usually absorbs blue and green wavelengths, very little light of any wavelength is available for plants to photosynthesise and there are almost no plants. Their place is taken by slower growing animals that usually live in deeper water. The buttongrass stained water lies in a layer at the surface and algae grow only to four metres. Below that, the water is clear but dark. On silty bottoms, sea pens spaced at regular intervals wave in the current. On the reefs, thin sea whips, many with basket stars waving from them, move as you pass. The 'living fossil' elephant shark and the more dangerous seven gilled sharks make the area unique and fascinating.

Divers in Adelaide like to dive Apollo Bay and offshore Kangaroo Island, Australia's third largest island. It is possible to transport car and boat on a trailer to the island by vehicular ferry but easier to fly there. One of the features of Kangaroo Island is the large number of wrecks that have come to grief here. Accessibility varies from the *Fanny M* which sank in 1905 on a sandspit off Kingscote in four metres of water, to the *Loch Vennachar*, which experienced divers can attempt when the big swells from the Southern Ocean are not creating too hazardous conditions close to the cliffs. Penneshaw Jetty and Reef are on the north east tip of the Island

Living in deep water between 20 and 70 m, the southern sea whip produces a flexible spiral which grows out into the open water where its polyps may feed. The jewel anemones are usually found on rock walls, but here they make use of the sea whip to reach the open water above the reef where there is little competition for the plankton in the water.

Diver explores a sponge garden in the cold waters of Bicheno, Tasmania.

facing the mainland at Cape Jervis. The Jetty is a safe dive and the wooden pylons have plenty to see growing on them, from ascidians the size of basketballs to gorgonia and sponges. Penneshaw Reef lies about thirty metres away parallel to the Jetty, but a strong rip over the top of the Reef when the tide changes makes it a slack water dive. Black cowries, crayfish and weed-eating fish are found on the Reef. Kangaroo Head is the site of Matthew Flinders' landing in 1802 in the *Investigator* to stock up on supplies of kangaroo meat, which is where the name comes from. West Bay, the site of the *Loch Vennachar*, though fairly inaccessible has magnificent underwater scenery, clear water, abundant fish and huge crayfish in ledges under the cliffs in about five metres of water. Western River Cove is sheltered from most winds by the cliffs on each side and further out, in eighteen metres of water near the headlands, are very tame grouper, harlequin fish, six-banded coral fish, blue devils and really big crays.

Closer to Adelaide are spots such as Port Noarlunga, a marine reserve with a jetty extending out to an offshore reef. The reef has the advantages of both accessibility and interesting marine life, and is protected and safe to dive for beginners and children, whilst retaining interest for more experienced divers with the presence of egg cowries, blue devil fish and gorgonia. Aldinga Reef, a marine reserve near Adelaide, is primarily a limestone rock platform but offshore is Aldinga Dropoff which provides interesting diving. Aldinga Reef is known for several white pointer shark attacks in the area. Boats can be launched at the north end of Aldinga beach and a further three kilometres north, is the Dropoff. It extends for about one kilometre, so the easiest way to locate it is to move inshore from deep water until the shallower part, around four to five metres, becomes visible. The Archways on the northern end in five to ten metres of water makes an interesting photographic site and the best diving is where it meets the sand in twenty metres of water. On longer trips, Adelaide divers like to visit the Gulf and Yorke Peninsula. Reef Head on the south western tip of the Peninsula, where ninety-metre cliffs drop straight into the sea, has an interesting cave system although the area is more popular with spearfishermen. Cave diving is, of course, a specialty in South Australia and the caves are world renowned.

Western Australia has the longest stretch of coastline of any Australian State, so some of the best diving is to be found there. It has coral reefs just off the coast, and shipwrecks dating back to the 17th century, so there is an enormous variety of underwater scenery. The more remote areas are virtually unexplored territory.

Perth has some spectacular diving right on its doorstep. Ferries leave Perth every few hours for Rottnest Island. There are a large number of reefs off the island where coral occurs, especially below ten metres where plate and honeycomb corals, gorgonia and brain corals are found. Tropical and temperate fish — butterfly fish, yellowtail kingfish and even cleaner fish occur on Transit Reefs, Pocillopora Reef, and the reefs north of Longreach Bay. A glass-bottom boat takes visitors to the wrecks on Transit Reef. In June, crayfish are found in large numbers as well as blue devils, buffalo bream, and fox fish that are present in large numbers all the year. Transit Reef is a graveyard of ships, the *Macedon* forms part of a shipwreck trail. Other ships are the *Denton Holme*, the *Transit*, *Janet* and the *Gem*. Visibility in Thompson Bay, which is the north eastern side of the Island, varies from ten to twenty-five metres. Armstrong Bay, on the middle of the northern side of the Island, is a reef that shelters a lagoon where unusual fish such as the pineapple fish and boxfish, blue devils and fox fish, congregate amongst the sponges, gorgonia and solitary corals. The Horseshoe Reefs off the north west end of the Island cover a large area and vary greatly in depth, the farthest about two kilometres out in a northerly direction in eighteen metres.

The deeper reefs offer some of the best big game territory at Rottnest with blue grouper, jewfish, samson fish, baldchin grouper and queen snapper more numerous than any other location. Pink snapper, which are caught by line fishermen in deep water, can be seen schooling and big crayfish are found on the ledges and in dark holes on the outer reefs. On the south west tip of Rottnest is Cape Vlamingh, known locally as West End, where huge pelagic fish school just offshore from the cliffs — large samson fish, Spanish mackerel, bonito and yellowtail kingfish. Seaward of the steep cliffs that surround the island and about one kilometre from shore are huge caves in twenty-five metres of water. Several are as big as churches. The caves offer protection to the reef fish from the strong surge that is present and small fish are very tame and follow divers, so that it is not uncommon to

The Gulch from Whalers' Lookout, a granite knoll overlooking the township of Bicheno, Tasmania and the Tasman Sea. The waters of the Gulch are clean and clear of pollution and are the home of a fishing fleet which supplies the two seafood processing plants.

Sugarloaf Rock in Western Australia's far south west, is a popular summertime dive location. The deep waters on its seaward face are home to a variety of local fish.

be followed by schools of footballer sweep, wrasse and leatherjackets. The fish may also be looking for a handout as divers collecting crayfish in the summer unwittingly leave behind a trail of legs and feelers, broken off whilst extracting crays from their ledges.

South of Perth and accessible from Busselton are Cowaramup Bay, which is suitable for shore diving and Canal Rocks for boat divers. Busselton itself has a two-kilometre jetty, once the longest jetty in the Southern Hemisphere. Jetty diving here is spectacular with soft corals towards the end of the jetty in over seven metres of water. Geographe Bay, on which Busselton is located, has several coral bommies that are Western Australia's southernmost corals. At Augusta, on Australia's south western extremity are many deep reefs accessible by boat from Flinders, home of the unique West Australian jewfish that come into the area when the water warms up. The fish is unmistakable underwater by its thick-set body, large eyes and mouth and is very inquisitive. Unfortunately numbers are declining, as they are very good eating and popular with fishermen and spearfishermen. Penguin Island, which is north of Busselton, is accessible by ferry or by walking across the waist-deep sand bar. Air is available at Rockingham four kilometres away. There are many reefs and islands north and south of Penguin Island that are the homes of seals, penguins, buffalo bream and scaly fins, with crayfish on most reefs and gorgonia corals very abundant and some stinging hydroid. East of the island on Five Fathom Bank is the wreck of the *Carlisle Castle*, an iron sailing barque, which sank in 1899, and is one of many wrecks on the Bank.

Albany on the south west corner of Western Australia is renowned because it was the site of the Cheynes Beach Whaling Company, until 1978, when it ceased operation due to declining catches and increased pressure from conservationists. The area is being turned into a museum of whaling that will include a whale chaser. One of the other whale chasers has been sunk to provide an artificial reef and dive site in twenty-one metres of water at the western end of Michaelmas Island in King George Sound. On the bottom with it is another whale chaser and coal carrier hulk. Albany is sited on the north side of King George Sound which faces in an easterly direction. The bay has a spit projecting from the south side about half way along, cutting it into two, and creating protection from the strong south easterlies. Diving within the protected sound and around several islands you can see relics of the whaling industry in the form of whale skeletons and teeth, and further out are such sites as Boulders, Michaelmas Reef and Eclipse Island.

North of Perth is Cervantes and in between the coast is dotted with small islands and reefs a bit off the beaten track but good diving territory. Jurien Bay is one such site 227 kilometres north of Perth, then thirty-nine kilometres west to the town. There is no ramp, just beach launching but the offshore limestone reefs, about two kilometres offshore, offer good visibility and a unique brilliant yellow tube worm which make it different from other reefs in the area.

Western Australia's Lake Leschenaultia is the only diveable freshwater lake within easy reach of Perth. It is set among bushland in the Darling Ranges, forty-eight kilometres east of Perth. The best entry point is along the dam wall, near the base of the jetties. The bottom slopes rapidly to four metres from the jetty and debris litters the bottom, then it slopes to twelve metres. On the bottom, mussel trails wind through the mud and gravel. Unlike their saltwater relatives, these mussels are not attached and must seek their food. There are small shrimp on the wooden pylons and a green freshwater sponge, but the

Vibrant colours of soft coral grace the reef edge.

Photographed at Sugarloaf Rock, the colourful harlequin fish sits in a sponge garden awaiting an unwary fish for lunch.

A diver descends among the pylons of the main jetty at Lake Leschenaultia, one of the few diveable freshwater lakes in Western Australia.

feature of greatest interest is the freshwater crayfish or marron, unique to the south west of Australia, which is found in hollow logs or under rocks. The lake is a dam, constructed in 1912 to supply water to a steam engine and is fed by five creeks which dry up in summer. The water is extremely cold and visibility may go up to ten metres in the winter when it is cold enough to take your breath away.

Western Australia, like other states, has its share of seal islands, but the one officially recognised is forty kilometres south west of Perth in Shoalwater Bay. The island is limestone with high cliffs on all faces except the eastern side which has a wide sandy beach, that is a big attractions for seals. The seals are New Zealand fur seals, and grow to 2.5 metres in length and weigh 180 kilograms or more. Six years ago there were rarely more than three or four seals, but other seals seem to be migrating there from seal colonies at Carnac Island, Little Island and Dyer Island near Rottnest. The numbers at these islands have been dropping over the same period. The seals head out to Five Fathom Bank, five kilometres west, to feed and perhaps this long distance feeding spot is the source of the migration. The seals at Seal Island will enter the water when the diver snorkels around the beach, and will mimic the diver blowing bubbles. They can be hand fed fresh fish from the boat. On the beach or in the water, hand feeding can result in a fight between seals and the diver may get involved. Off Seal Island, 200 metres west is a reef with many caves and ledges for crayfish, gorgonia and fish, where the blue devil, with his iridescent blue spots and yellow eye, is common. Seal Island is not far from Penguin Island.

The coastline waters of southern Australia offer some spectacular and unique sights. In this chapter we have explored only a few of the more popular sites. There are many more wrecks and reefs to be discovered by the adventurous diver exploring off the beaten track. In the following chapter, diving is discussed in the inland freshwaters, where divers can explore the drowned limestone cave passages, one of which is the longest cave dive in the world.

The swimmer floats suspended on the surface as the sun's rays illuminate her.

In coral seas, large caverns are often found inside small entrances.

CAVES AND CAVERNS

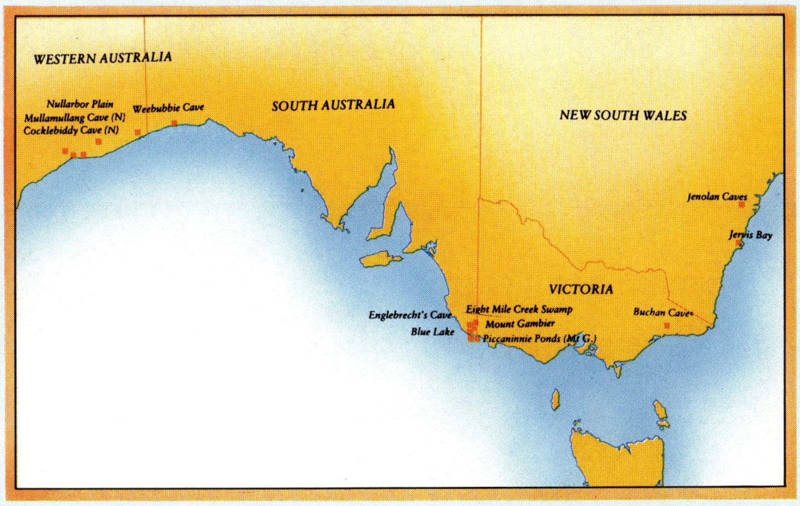

Whether along the edge of Australia's coast, within the large piles of granite boulders on the sea floor, or one of the many sinkholes and underwater passages inland, caves and caverns are a source of fascination for many divers.

There are three main types of caves in Australia: weathering caves which include shallow caves and rock shelters that are caused as a result of the atmospheric weathering of rocks such as sandstone; sea caves that are found along the coastal cliffs of the temperate southern waters and are caused by powerful wave action; volcanic caves which were formed in lava several million years ago; and limestone and dolomite caves, the largest and most complex of Australia's caves, which were formed in periodic phases over millions of years ago.

All types of caves provide special habitats for animal life which are of great biological interest both ecologically and for the study of evolution of species. Sea caves offer protection for marine life whether it is for the hundreds of fish who school in coral seas or for the large colonies of seabirds and seals that seek sanctuary in the stacks and arches on the southern coast. In many limestone caves, bones and fossils have been found and these have helped to assess the age of the cave and to indicate its former surface environment.

Most Australian caves are in reserves or national parks. In New South Wales the best known cave systems include the Jenolan Caves and those in the vicinity of Jervis Bay. In Victoria, the sea caves of Buchan, Lower Glenelg and Port Campbell are well known as are the caves of the Mt Gambier region in South Australia and the caves of the Nullarbor National Park, which contain fossils of extinct fauna, and the series of limestone caves in the south west of Western Australia.

"In the calm winter months I dive in Smugglers' Cave on the northern headland of Jervis Bay. We anchor on the seaward side and enter the water. The cave entrance is in ten metres of water, but extends both above and below the water with the surface accessible directly above me at the start of the dive. As I go into the cave, it opens up underwater. Further along the airfilled chamber above me terminates, but underwater the passage continues. I reach a narrow pinch and the sound of the waves pounding above hits me with a compression force, and thunder rings in my ears. I feel that I am breaking the sound barrier as I swim under the ledge through the narrow pinch and enter the second part of the cave. I see sea spiders adorning the gorgonia that lines the walls. A wobbegong flies off the bottom as I disturb him. As I swim out into the light, a feeling of awakening or rebirth exhilarates me."

The dive boat is at the entrance to one of the large sea caves at Jervis Bay. The sandstone cliffs have weakened along bedding planes and vertical faults, so that great square blocks have fallen into the sea creating caves with square entrances.

An appreciation of coastlines and their formation introduces this chapter because caves are a feature of weathering of the coast and hinterland. The coastline of Australia extends for 36 800 kilometres and ranges from tropical climates of Cape York to the cool temperate climate of Tasmania. The coast is evolving as a result of numerous geological and climatic influences. These factors vary from State to State as the climate varies, and depend on the structure of the rock composing the coast and hinterland.

Climate influences coastal features in a number of ways. In the tropics, humid conditions accelerate chemical weathering, destroying many surface features rapidly. Bold headlands are found only where there are resistant rocks. Alternate freezing and thawing in colder climates shatters the rocks which are eventually sorted into beaches and shingle. On tropical coasts an abundance of fine sediments is carried to the coast from the hinterland by rivers. Mangroves stabilise these sediments building up the land. In arid areas very little sediment is transported to the coast and beaches consist largely of sediment derived from biological inhabitants, coral and shell debris.

Wind generates currents and waves, which create water circulation patterns shaping the sea floor, moving coastal sands and shaping dunes on the foreshore. Tides and salinity affect weathering and transportation. Time is an important factor — many of the features we see are the result of processes that occurred during the last Ice Age, when ice sheets covered the Poles locking up sea water and lowering sea levels, which rose again as the temperatures increased.

The animals and plants that live on the coast both above and below the tide level are determined by climate. In turn they influence erosion and building processes. Corals and reef building organisms inhabit the tropics and subtropics, creating coral fringing reefs and providing calmer conditions on the coast. Mangroves grow in estuaries and sheltered coastal environments in the tropics, while in temperate areas, salt marshes replace them. Marine worms boring into coral break it down and compact sediments on coral reefs, aided by coralline algae.

In temperate waters creatures, such as the bivalve *Barnea*, bore into rock. It uses its shell valves as it twists or rocks on its foot. Sea urchins may also aid the break down of rock — helped by an acidic substance they produce that enables them to create cavities in the rock.

The seal from the colony on Montague Island is caught in the rays of the setting sun. One must be wary of the large bull seals in the colony which may be aggressive.

Sea caves most commonly occur on cliffed coasts where the impact and withdrawal of waves abrades the cliff face. They are a feature of temperate Australian areas. Rocks that are most resistant form the rocky headlands and offshore islands, while the less resistant rocks are cut back as coves and embayments. Sandstones and rocks with closely spaced structural weaknesses or bedding planes, erode more easily than massive formations of resistant rocks.

Natural arches form where the less resistant rock has been excavated from a promontory, leaving an arch such as London Bridge in Port Campbell National Park. When the hydraulic action of waves against the roof of a cave makes a cavity, eventually blowholes such as at Kiama, New South Wales, occur. When the roof of the cave collapses, steep sided gorges result such as at Jervis Bay, New South Wales, and Port Campbell, Victoria. Natural arches and stacks such as the Twelve Apostles at Port Campbell occur where powerful wave action from the southern ocean has cut out ledges in rocks at the levels of the bedding planes. These rapidly changing coastlines are the result of recent geological processes.

At Jervis Bay, New South Wales, caves dot the huge cliffs with square shaped mouths where vertical faults and horizonal bedding planes have combined to form square shaped entrances. Smugglers' Cave penetrates 300 metres into the headland and, after a sharp right hand turn, it exits on the other side. One of the recently discovered caves in the area has an underwater entrance, suggesting there may be more awaiting discovery.

The height of cliffs is related to the distance of uninterrupted travel by wind-generated waves. Thus cliffs facing the ocean's full force are much higher than those on protected coastlines.

Where coastal rock formations do not have great structural variation, intertidal rock platforms may have developed. If the rock type is resistant, they may not have developed in the time since the sea reached its present level. A high wave energy coast, where rock debris aids wave action in abrading the foot of the cliff face, is more likely to have rock platforms. Intertidal rock platforms form by a process known as 'water layer weathering'. Salt spray and rainwater erode exposed rock surfaces above high tide level, honeycombing the rock. The rock surface disintegrates above the level of permanent saturation by the tide and peels down to a level that remains intact because it is permanently saturated. Such horizontal platforms may traverse local structural rock features. In more resistant rock, such as the granite and quartzite at Wilson's Promontory, weathering is less effective and high tide shore platforms do not occur. On limestone coasts, low tide shore platforms develop that are broad and flat, the cliff face often overhung with a visor, and the outer edge forming a higher rim of encrusting algae. Although the seawater off limestone coasts is saturated with calcium carbonate, a nocturnal cooling and increase in acidity allows it to dissolve more limestone within the tidal zone, at night. This type of platform develops on the southern and western coasts and with some variation on tropical coral coasts.

Huge boulders often fall into the sea from coastal cliffs, decorating the sea floor like giant marbles. They create underwater caves, caverns and swim-throughs. One such boulder at Jervis Bay is hollow. Divers can swim inside where bubbles from a diver's exhaust collects in an air pocket twenty metres underwater inside a boulder. It is fascinating to talk to your companion inside such an air bubble deep within the sea. These bubble caves, as they are called, are also formed naturally by wave or surge agitation, collecting in mirror pools in the roofs of underwater caves.

Coral creates caves underwater too. Large caverns become overgrown at the entrance, where coral is growing in the light. Inside, where light levels are low, hard corals are absent but are replaced by soft corals and gorgonia. The soft corals may enlarge many times their size in the late afternoon and evening when they feed, puffing up as they filter feed and expelling the water again during daylight. These coral caves may have other entrances almost overgrown with coral, but used by fish escaping predators and by divers as windows to see outside. In more temperate waters, sunshine anemones line the cave walls as in Fish Rock Cave at South West Rocks on the New South Wales north coast and Solitary Islands off Coffs Harbour, which also features such 'fields of flowers'. On coral islands, ocean caves often occur at twenty metres. It is at this depth, where coral is actively growing, that overhangs and coral caverns develop. In coral tunnels greater water flow inhibits the growth of large hard corals. Soft corals and smaller species replace them. Butterfly cod, a poisonous stinger if touched, often lurk in these caves.

Plunging cliffs, pass steeply below water without any platform, and may result from faulting, like a slice with a knife, in which the cliff is the vertical plane of the fault and the horizontal block, the other slice, lies beneath the sea. At the Docks, Jervis Bay, the headland has been faulted and eroded forming a sequence of underwater clefts and platforms similar to a series of Docks

The entrance to Smuggler's Cave, on the northern headland of Jervis Bay.

The wobbegong shark is a common resident of southern waters, preferring the gutters caves and sponge gardens. Large numbers are present in the Jervis Bay caves. They move quite quickly when feeding and eat octopus among other things. Divers have been bitten when handling them in the mistaken belief that they have no teeth. Since they are capable of turning on their own tail, it is not very sensible to approach them from behind and tug their tail, but if approached carefully there is no reason to fear them.

that a small ship would berth in. One of the smaller ones is thirty metres long and two to five metres wide with the surface visible twenty metres above.

Volcanic activity can account for another type of plunging cliff where there has been insufficient time for shore platforms to develop. Lord Howe Island has volcanic plunging cliffs with irregularly shaped caves. Very calm seas are needed for exploration of these caves as large ocean swells can smash divers or boats against the back wall of the cave.

Storm activity on the Nullarbor coast is abrading these southern cliffs. They extend to a level below low tide level and so form cliffs that plunge into the sea.

Caves offer protection to marine life. Schools of fish can been seen orienting to the nearest wall. Species that are usually found in much deeper or darker waters, such as gorgonians and black coral, grow in caves while deep sea fish, such as the pineapple fish and the angler fish, may be found hiding there. Light is a major controlling factor underwater. Sunlight is absorbed progressively with depth influencing the growth of algae and animals such as corals that are intimately dependent on the zooanthellae algae in their tissues. The light levels in caves are equivalent to those found at much greater depth, hence many deep water inhabitants favour caves for part of their life. Crayfish often shelter in caves, and it is not uncommon to see hundreds of crays covering every ledge and cranny in a large cave, especially during winter. They come out of the cave to feed at night and hide during the day to escape predation. Double-ender or slipper crays and tropical painted crays

The Cathedral in Piccaninnie Ponds has a narrow entrance which Chris Deacon has just entered. Inside the dark cathedral, no algae grows and the sculptured limestone walls reflect the diver's lights.

like caves too. Large schools of morwong, weighing up to 2.5 kilograms each, inhabit the deeper caves in temperate Australia. These edible fish are safe here from fishermen's nets. Coral-banded shrimp favour the cave entrances, where they clean parasites from fish that come to these 'cleaning stations'. They are found in tropical and temperate waters.

Bigger animals shelter in caves too. Wobbegong sharks often lie on the sandy bottom of small caves, their camouflage colours making them almost invisible. Port Jackson sharks can be handled very carefully when they congregate in caves to mate during winter. Later their corkscrew-shaped egg cases can be seen among the debris and boulders outside the cave or among the seaweeds washed up on the shore. Stingrays, up to two metres across, may be buried in the sand of cave floors, with just their eyes visible to the observant diver. The blue devil wrasse may be spotted in the darker regions of caves with the aid of a torch. Nannygai mill in caves with bullseyes, old wives and other fish.

The smaller cavities are homes to moray eels, looking more ferocious than they are, they will often dart out to take a look at passers by. They can be lured from their holes with a piece of fish or abalone.

Besides the sea caves of Australia there is another type of cave that divers explore: the freshwater limestone cave. Limestone occurs in many places in Australia, decorating caves with stalagmites and stalactites. The drowned water passages and sinkholes that divers like to explore are not so common. Sumphole divers have explored Jenolan Caves in the Blue Mountains, New South Wales, Yarrangobilly Caves in the Snowy Mountains, Buchan Caves in Victoria and caves in Tasmania that extend 1300 metres, but it is the Mount Gambier area in South

Lord Howe Island is volcanic and caves formed in volcanic rock are irregular as can be seen here. Dive master Jeff Deacon is manoeuvering his dive boat into the cave so divers can explore it. Calm seas are needed for this type of diving — large ocean swells can smash the dive boat against the back of the cave. A large swell entering a similar cave at Jervis Bay caused a fatal accident to divers who were driven into the small ledges.

Caves sometimes have small entrances called windows which allow light through and the play of light on the floor of the cave creates a lovely effect.

The guide diver is using a cave line and reel and the group following are clipped to the line as they enter 1080 Cave. The lead diver will exit as he reels up the line.

Australia that has attracted the most attention from divers. A period of volcanic activity intruded along faults and joints in the limestone of the Mount Gambier region and it was during this period of subterranean activity that the complex of four lakes of Mount Gambier formed. There may have been two eruptions: one 4830 years ago; and another about 1400 years ago. Aboriginal legends relate to Mount Gambier as an 'oven' that 'thundered into the ground'. Earthquakes in 1897 and 1948 shook the area.

During the last Ice Age, seas were lower as the result of the withdrawal of water into the Ice Caps at the Poles. A series of wind-blown dunes and marine terraces were formed along the coastline that now mark fluctuations in sea level.

Mount Gambier's most famous feature is the Blue Lake. The water in the Blue Lake derives mainly from a confined aquifer or closed system that is recharged with water from areas to the north of the town. The lake is a water-filled volcano that changes colour overnight late in November each year from its sombre winter grey to a bright turquoise blue, a colour it retains until March. The Blue Lake has a five-kilometre circumference and a surface area of twenty-eight hectares and its water surface is seventeen metres above sea level. The Blue Lake is Mount Gambier's water supply. The confined aquifer is the major source of water in the Blue Lake but somewhere underground it mixes with water from an unconfined aquifer and this may be a cause for concern as pollution has occurred in the latter aquifer. This unconfined aquifer occurs in rocks closer to the surface from about two to thirty metres below the ground. This groundwater moves over a low slope towards the coast discharging in springs such as Ewens Ponds. It moves through a well developed trough from eight kilometres west of Mount Gambier through Mount Schank, under Barnoolut Station (site of several diving sinkholes) to Ewens Ponds, where it discharges from Eight Mile Creek. This trough is aligned with regional

Above: Chris Deacon glides effortlessly through the canal joining the Ewens Ponds where there is a slight current.

Below: Divers about to enter one of the freshwater sinkholes in the Mt Gambier area.

Little Blue Lake at sunset, the sky reflected in the still waters of the sinkhole.

Bullock's Hole has a fairly narrow entrance. Erosion of the land has created an immense cave inside. Special techniques are needed to get divers to the water and ladders must be attached to a vehicle to enter the hole.

geological structures and is the preferred flow path of the water in the unconfined aquifer.

This unconfined aquifer is recharged by water percolating through pine forest and pasture. Sinkholes and drainage bores provide further recharge as do the paved urban areas of Mount Gambier, though the latter is an insignificant amount. It is also recharged from the old dune formations that occur along the coast. Before 1963, all Mount Gambier's waste water drained underground, but a sewerage system has now been installed. Very high nitrate concentrations have been measured in Mount Gambier's water supply. The normal safe limit for drinking water is forty-five milligrams per litre and concentration ten times as high or 490 milligrams per litre have been measured. Since the naturally occurring nitrate in the groundwater is very low, this high concentration of nitrogen represents pollution. Two sources are responsible: city effluent disposal accounts for 1.5 milligrams per year which, over 100 years of disposal, accounts for 25 to 100 milligrams per litre; the other contributor to the high nitrate levels is stock (cattle and sheep) that are thought to add about 1.5 milligrams per litre.

Bacterial pollution is also evident and some stock watering bores have been closed for this reason. Phosphates that derive from detergents and superphosphate fertiliser are not significantly high, which is fortunate since phosphates are associated with nitrates in the production of algal blooms and associated hydrogen sulphide or 'rotten egg gas'. Discharge of water from the unconfined Mount Gambier aquifer is through evaporation and groundwater flowing east through the Glenelg River as well as through Eight Mile Creek. Changing the land use affects the recharge and discharge of the system. Recharge of aquifers in pine forest is less than on pasture. Farmers use bore water for irrigation causing greater loss through evaporation. While loss of water from the system is of some concern, losses caused by man's interference account for only twenty to twenty-five per cent, the remaining seventy-five to eighty per cent is natural discharge.

Management of the unique resource of Mount Gambier in terms of both its water supply and tourist potential is essential. Fortunately it is not too late to take action. Water in the Blue Lake has not been irreparably contaminated by nitrate because of the slow rate of movement of the groundwater and the contribution from the confined and unpolluted lower aquifer.

Siting of bores and correct construction methods, including locating stock troughs at least 100 metres from the bores, is one way to minimise nitrate pollution from stock. Waste disposal legislation is only effective if the laws are imposed on both public offenders and private individuals, so that unauthorised pollution of drainage bores does not occur from farms or sewerage works. It is also important to see that the water table or water levels in the system do not drop from uncontrolled or increased use. A drop in the hydraulic head could cause leakage of polluted water into the confined aquifer causing pollution of the Mount Gambier's water supply and the unique tourist resource — the limestone caves.

The sinkholes of Mount Gambier are one of Australia's famous sites. In the early 1960s divers began exploring the sinkholes around Mount Gambier near South Australia's eastern border. Piccaninnie Ponds was one of the last holes to be discovered in 1962. A spectacular pond at the back of the coastal dunes, Piccaninnie Ponds is one of Australia's best diving locations. It is set in a 600-hectare national park administered by the South Australian government. The salinity of the Ponds is higher than in other sinkholes in the region and this has modified the plant life found there to the hardy species — water milfoil, moss and algae — that occur to a depth of sixteen metres. The swamp originally drained into the Glenelg River to the east but a channel (now overgrown) was cut into the sea on the western end increasing outflow from the swamps and lowering the water level. This was probably done to increase grazing land or to improve the habitat of the short-finned eel that inhabits the Ponds.

Bulrushes, sedges and typical bog plants grow on the Ponds' fringes, while water milfoil, a plant not present in the nearby Ewens Ponds, grows at the edges of the deeper chasm with filamentous algae entangled in it. The aquatic moss grow in clumps on the walls. This species is usually found on forest floors, but here it occurs on the limestone walls down to sixteen metres. It had not been recorded from underwater until recently when both it and another moss were found in the Ponds. Blue-green algae, a primitive plant group, grow in purple clumps beside it, contrasting with the green moss.

The food chains in the Ponds begin with digestion of organic detritus and silt by foragers. There are diatoms living on all the water plants in the Ponds and it is thought that these obtain nutrients from the plants which in turn absorb nutrients from the rich peats overlying the calcareous sand and limestone in which they are anchored. These plants support a complex ecosystem that includes the eels, turtles, mature trout and even sea fish and yabbies that live in the Ponds. The protection of this fragile freshwater ecosystem has been ensured by the declaration of the Piccaninnie Ponds National Park. A ramp has been constructed to allow access to the Ponds without damaging the swamp vegetation. A permit is required for both scuba and

Above: Divers preparing to enter the water in 1080 Cave (named after the rabbit poison) at Mt Gambier.

Middle: Cave diving instructor Ian Lewis leading a group into Piccaninnie Ponds. This short, shallow penetration requires no guideline if divers stay away from the walls.

Below: Divers about to enter the water at Goulden's Hole. The landowner's irrigation pipe provides a useful orientation point for the dive team.

The sheer walls of some sinkholes make planning essential. Equipment must be lowered and divers require a landing to gear up. Sometimes divers gear up on the lip of the sinkhole and clamber down a ladder while the support crew lowers scuba gear to the diver in the water. Exertion at the end of the dive while waiting for equipment to be retrieved by the support crew can bring on decompression sickness, so it is essential that these dives are well planned.

snorkel divers. This precautionary measure of making permits mandatory was taken after two divers lost their lives in the Ponds, one in a shallow cave accessible to snorkel divers. This is a Category Two Cave Dive, as discussed in cave diver training in the first chapter.

The excitement of Piccaninnie Ponds is best captured by this description of one of my dives:

> Water gives me that feeling of weightlessness of falling or gliding to the bottom. Algae, growing like forest trees, surround me as I descend between them, touching their silky strands. I look back to see the others drift through the forest in dreamy unreality. Clear white walls of sculptured limestone contrast with the dark foreboding hole that is the entrance to the cathedral. Past it, the chasm drops to over sixty metres — past the limits of sports divers, my reality warns me. I enter the cathedral and am filled with awe as it opens before me.
>
> The cathedral channels off under a ledge and I am tempted to explore deeper water, but I am content to shine my torch into the blackness. Later I am in a group of divers who are exploring Piccaninnie Ponds at night and we see the stars as we look up through the water and I feel myself falling on my back slowly to the bottom. I can see torchlight below as the divers enter the cathedral.

Piccaninnie Ponds is an experience that only divers can enjoy but one day it may be accessible as are Florida Springs and other freshwater environments. A submersible chamber would allow visitors to see one of Australia's natural wonders. Film crews from all over the world have dived in Piccaninnie Ponds and presented its wonders yet few Australians know and appreciate it.

Ewens Ponds are a series of springs set in the coastal scrub along the road to Port MacDonnell. The Ponds empty 2.3 cubic metres of water per second into Eight Mile Creek and thence to the ocean. The sand boils at the bottom of each pond where this huge volume flows out of the ground. Entry to the Ponds is from a ladder off the jetty. It's a great thrill to ride the current in the narrow linkways between the Ponds. There is a ladder on Pond Three where you can get out of the water or you can buzz the full distance of Eight Mile Creek to the sea. Until spring 1984, you did not need a permit to snorkel in Ewens Ponds but since then, some restriction has been placed on large numbers using the Ponds in peak periods as there has been some degradation of the plant life.

The Eight Mile Creek swamp, so called because it is eight miles from Port MacDonnell in South Australia originally extended along the coast between Nelson in Victoria and Port MacDonnell and for 3.5 kilometres inland. Changes in sea levels since the last Ice Age have left a series of ridges representing old coastal dunes. The Eight Mile Creek springs were formed when faulting released water from the unconfined

In Piccaninnie Ponds, algae grows in the shallows where light is available. It extends from the depths up to the surface here in the chasm leading to the Cathedral. Divers have not reached the end of the canyon.

Diver's light reflects the algae growing like a forest from the cave floor in one of Mt Gambier's freshwater lakes.

Farmers sometimes use the freshwater sinkholes for stock watering and in this case have provided an access ramp to the water which divers use.

aquifer resulting in a swampy area. This area was drained for farming land with the result that the water level has dropped by 1.5 metres since 1947, and the area of swamp has been reduced. Before drainage, the dominant plant species was watercress but it is now restricted to the lower reaches of Eight Mile Creek below the Ponds. Water plants such as milfoil were once found at Ewens Ponds but are now restricted to Piccaninnie Ponds. Swamp species of reeds and sedges surround the Ponds, but again there are changes in the dominant species and loss of some species as a result of the changes of land use in clearing for pasture. The water plants help stabilise the slope below water level, where there are inclines of forty-five to fifty degrees. Below two metres, blue-green algae stabilise the very fine organic layers and calcareous sands, where the root systems of the underwater flowering plants are unable to take hold. Apparently, the flowering plants can self-pollinate underwater, but it is thought that the fruits cannot dry out and release seed so they reproduce asexually without pollination. The dense growth of bog plants around the Ponds ensures the Ponds remain clear even after heavy rain, as there is very little runoff into the Ponds. The limestone aquifers have an enormous filtering capacity and the rapid turn-over time of the Pond water varies from 1.5 hours in Pond Three to six hours in Pond One. Around the Ponds, bird and insect life abounds and there are eels, salt water fish and the rare freshwater crayfish, called prickly-backs, that carry flatworms or caterpillars which build cocoons on the weed.

Ewens Ponds is a Category One dive with no tunnels or overhangs where divers can become lost, and the surface of the water is directly accessible at all times. In late afternoon or on a moonlit night, when the sky and stars are visible above, it appears there is no water around you at

Chris Deacon swimming along the waterway joining the ponds where beautiful water plants climb towards the light.

Above: Warning sign at Little Blue Lake which is accessible without entry through private property.

Below: Chris and Kevin Deacon about to enter the water at Ewens Ponds by the access ladder that prevents disturbance to the environment.

all and Ewens Ponds becomes a magic place where you might expect to see a leprechaun around the next bend.

Mount Gambier's Blue Lake is an extinct volcanic crater formed when Australia passed over a hot spot in the Earth's mantle. It is one of Australia's younger volcanoes. It is in a complex of lakes that includes Blue Lake, and Valley Lake which is accessible to boats, and the smaller Brown Lake and Leg of Mutton Lake.

Little Blue Lake is a local swimming hole easily accessible down a ramp along the road to Port MacDonnell. The surface water is generally poor in visibility but as you go through several temperature layers or thermoclines to the bottom, visibility increases. When you get to the bottom, it is covered in thick silt and rubbish. Although it is a Category One dive, a guideline is necessary in the poor visibility. Other Category One dives in the Mount Gambier area are The Sisters, Horse and Cart Sinkhole, Tea-Tree Hole and Gouldens Hole. The latter is named after a previous landowner, and used for irrigation via a pump shed at the bottom of an access ramp.

Black Hole is a very large sinkhole that was called the Devil's Punchbowl. The overhanging sides drop ten metres to the water. Divers use a ladder to enter the water. Since there are no trees near the hole, ladders must be attached to the vehicle. Once inside, the huge cavern opens up beneath the land. The reflection of the tiny entrance is visible through the water. The lake is usually green with algae, but below the thermocline or temperate layer, visibility improves and a tunnel leads to a depth of fifty-one metres. Both this cave and Ten-Eighty Cave are on Barnoolut Station. Permission for these Category Two dives must be obtained by writing to the landowner two weeks in advance. Ten-Eighty is named after the rabbit poison and has an eerie greenish light, the result of algae in the water in the upper layers. From the central rockpile in Bullock Hole, also on Barnoolut, but a Category Three dive, you can see people leaning over the small entrance through thirty metres of water and air. Ela Elap is an Aboriginal name meaning small fish. This sinkhole has a ramp with easy entrance near the pump shed. Visibility is poor near the surface but improves, though the water temperature at eleven degrees is very cold for Australian caves.

Some sinkholes in the Mount Gambier area are on Forestry land. Hell's Hole is a Category Two dive with a thirty-metre ladder straight into the water. Divers must beware of falling rocks, as well as snakes which fall in and live on the ledges at water level. A climb down the ladder is required and a support crew to lower the tanks to the divers in the water. It's easy to see how the cave got its name.

The Pines is another Forestry sinkhole with a large chamber but only a small lake entrance in the pine forest. Shafts of light like laser beams penetrate the clear water, making the pine forest visible through the surface in the late afternoon. This is a Category Three dive as the lower tunnel is confined and silty.

Allendale Sinkhole is known as Middle-of-the-Road, because that is exactly where it is, right in the centre of Allendale across the road from the store. An introduction to Category Three diving, the chamber is large and the access is down a slope of rockfill placed there from earlier attempts to fill the fifty-one metre deep hole.

Fossil Cave is alongside the road to Millicent and, from the top, does not look diveable. The first fossils were found in 1964. Divers on several expeditions brought to the surface skulls of extinct kangaroos. A small pool at the south east end is the entrance to a graveyard of fossil remains. There are skeletons of now extinct animals that fell into the hole during a time of lower sea level about 17 000 years ago.

In 1979 Flinders University and the Cave Divers Association of Australia began an archaeological dig with a grid system laid out underwater that is still there today. The dig was conducted in zero visibility and divers had to feel their way through fine silt, their only reference point being the rope grid that took forty hours to set up on the bottom. They found complete skeletons of what were previously thought to be giant kangaroos but proved to be only a metre tall. Remains of extinct marsupial mammals: bandicoots, wallabies and rats as well as bats and birds were found. The fossil finds gave scientists a much better understanding of the life that existed in the Mount Gambier area 17 000 years ago.

Max's Hole is hidden under a windmill tower and a twenty-metre ladder climb makes it a dive for the intrepid explorer. It is a Category Three dive in very silty conditions. Category Three dives are in sinkholes and caves with submerged passages and silting conditions. Training to cope with conditions such as complete black-out are essential before diving in them. Silt may also result from dissolved lime lying in white powdery silt layers on the cave floor, walls and ledges. The bottom of the Pines is a good example, where white-out conditions can occur with the flick of a fin. In learning the techniques of cave diving, minimising the risk of disturbing silt is combined with the use of guidelines and other equipment and techniques. Tie-offs of the line may be required where there is a bend or obstruction. Buoyancy must be controlled to avoid stirring up silt on either floor or roof. Conditions in Australian caves are somewhat different from other parts of the world and different techniques are sometimes used. For example, floating guidelines that keep divers away from silt on the cave floor. In Florida, heavy guidelines that sink to the bottom are used, since strong currents are present in Florida caves, which are not present in Australia.

The restriction of entry into the South Australian caves is the result of a number of deaths including the loss of four divers in one incident in 1973. The Shaft, as the sinkhole is called, has been closed since then.

The first of Mount Gambier's caves to be explored was Englebrecht's Cave in 1864. Dr Wehl, a Mount Gambier physician, explored the cave with another three men and a boy. They carried a canoe into the cave. It was built of hardwood and sealed with pitch, and it took its occupants towards Blue Lake, the current making it difficult to get back, so the contemporary reports state. As the report in the local paper was written by the boy, perhaps it was rather a fanciful account, for no current now flows through the cave. The cave was named after the owner of the local whisky distillery. Brewery rubbish and refuse littered the cave until 1979, when the Lions Club of Mount Gambier began restoration. The entrance is in North Terrace, Mount Gambier. The cave was investigated in 1964 by divers to assess its tourist potential. From the shore of the lake, there is an underwater passage to an inner air chamber. It is the most advanced dive in the Mount Gambier area. The entrance to it from the underwater passage is narrow and not easily seen from below. The presence of an inner air chamber, reached only by swimming through a tunnel, is a preview of the next cave diving experience.

The Nullarbor is a desert plain centred on the South Australian, West Australian border. Peppered with saltbush and bluebush, it stretches as far as the eye can see from the sheer ninety-metre plunging cliffs that meet the Southern Ocean, inland for 250 kilometres.

The deep caves of the Nullarbor occur between the Hampton Plateau and the coast, a portion of the plain that is better watered receiving about twenty centimetres a year compared to half that further inland, while in past times much wetter conditions prevailed.

The Pines is a small sinkhole in the forest with duck weed floating on the surface of the water. The roots of pine trees can be seen on the ceiling of the cave and in one area a pine tree has collapsed through the ceiling and lies on the floor of the cave which slopes rapidly into a narrow and complex penetration dive.

Above: An exterior view of Cocklebiddy Cave on the Nullarbor Plain with a vehicle on the rim.

Below: Preparation for penetration diving is extensive, requiring a good support team.

Collapse holes or domes dot the surface of the Nullarbor on the wetter parts of the plateau within sixty kilometres of the coast and may be up to thirty-five metres deep and 240 metres across. There are about 120 of them and in travelling across the plain by vehicle, they present a real hazard because they are not visible until you arrive on the rim. Caves commonly lead off these depressions and, after rain, streams flow into them. The caves occur down to a depth of 106 metres from the surface and are weathered and fractured, their floors obscured by debris. The deeper caves also extend greater distances horizontally: the longest, Mullamullang Cave, has ten kilometres of passages. At intervals collapse domes occur, their ceilings up to sixty metres above the cave floor level. The deeper caves tap the groundwater and it is these caves that interest divers.

The horizontal passages, or chains of halls, in these caves suggest a geological formation along joints or weaknesses that are evident in the roof at Cocklebiddy Cave, and correspond with ridges evident on aerial photographs.

At a time when the groundwater level was about fifteen metres below present levels, ground water moving towards the coast began to dissolve the limestone. The present large cave passages formed when the roof collapsed after withdrawal of the water. A rise in water level then drowned the lower parts of the rockfalls piled on the floors of the caves. Groundwater has not caused solution of the cave debris, as the angular shape of the boulders attest. In Mullamullang Cave the solution passages are a metre or two in height and occur at higher levels. These are believed to be an earlier stage of cave formation, formed at a time of higher water level.

Aboriginal people knew the caves: the Kukuda, Mirning and Pindini people lived here long before Europeans. At Koonalda Cave they quarried flint and chipped patterns in the Art Passage, which is now in total darkness. For over 20 000 years they have lived in the Nullarbor and in that time the climate has changed from a milder wetter climate with greater abundance of food to the desert we see today.

Eleven of the caves on the Nullarbor have water in them. When discovered by pastoralists in 1901, they were used to water stock, even though they are salty. Two caves Weebubbie and Cocklebiddy have become famous all over the world, a fact that would certainly have surprised the first explorer of the caves, Captain Maitland Thomson, when he ventured into Weebubbie Lake in a three-metre wooden dinghy in 1851.

In the 1960s divers began to explore the caves, at first snorkelling in the lakes, then penetrating the underwater passages with scuba equipment. The remote area, and need for sophisticated diving equipment and techniques, had kept them isolated. Then, in 1972, seventy years after they were first discovered, Ian Lewis of the Cave Exploration Group of South Australia, organised a major expedition. It consisted of six divers and thirty cave explorers who were the support party. All the diving gear had to be carried over boulders and rough ground before reaching the water. Cocklebiddy Cave drops ninety metres over 200 metres horizontally. Other caves have a long sheer drop into the entrance chamber and rock climbing gear must be used to lower equipment. During this expedition and another two years later, they explored all the caves with water passages.

Other teams of divers began to explore the long drives of various caves reaching the limit of Tommy Graham's Cave and reached a point in Weebubbie Cave where depth and distance created problems. Cocklebiddy Cave attracted more attention, as the other caves were found to be dead ends.

Cocklebiddy Cave is a long shallow cave which, in 1973, had been explored to a point 550 metres from the underground lake. In the following year, a joint South Australian – West Australian diving team organised another major expedition. Using sophisticated equipment such as submersible radio direction finders, enabling accurate pinpointing of the divers' locations, and more sophisticated wet and dry cell batteries they were able to extend the cave limits further. A big advantage was gained by laying high pressure copper tubing from the cave mouth into the main chamber. This enabled scuba tanks to be refilled without being hauled to the surface, an enormous saving in time and effort. They reached a point one kilometre by underwater tunnel from the entrance lake, where a rockpile blocks the water passage. On the other side they discovered yet another lake and passage which they explored underwater for a further 150 metres.

In 1977 divers explored 2000 metres of passage, a return swim of four kilometres. This was

Below: The frantic preparation for the 'push' to the Rockpile is visible in this time exposure shot of the divers' helmet lights.

Diver in the maze in Tommy Graham's Cave with the sculptured limestone which is sharp edged, showing minimal water weathering.

possible using triple scuba tanks on their backs and pushing sixteen backup tanks into the inner chamber on an underwater sled. A base party set up camp on the inner rockpile after moving to there all food, sleeping bags, batteries and clothes in waterproof containers and setting up a phone connecting them to those outside the cave. Two Australian teams followed in 1979, one exploring three kilometres from the entrance. This took five hours by a team of nine divers with fifty-eight scuba tanks. It established, for Hugh Morrison and his team, a new world record, and for Australia, the longest submerged cave dive in the world.

They had still not reached the limits of Cocklebiddy Cave and in 1982 Hugh Morrison organised another expedition. An underwater sled transported the scuba tanks to the rockpile which again formed the base camp. After dismantling the sled and carrying it and all the tanks to the far side of the rockpile, the three divers entered the water.

They planned to use the one third rule of cave diving. When one diver had used a third of his air in the fifteen tanks on the sled, they left the sled with their return air supply and continued on their way, using the air in the triple tanks on their backs. Thus at 1000 metres from the entrance, they parked the sled on the roof of the cave. They passed the previous two-kilometre point and discovered another air chamber — Toad Hall. After three hours and fifteen minutes underwater from the rockpile they rested for an hour, in Toad Hall. They were concerned that those waiting at the base camp would think they had run out of air, so they began the return journey. It took an incredible total dive time of seven hours. Each diver had swum seven kilometres and physical and mental fatigue was evident as they emerged from the cave to a Nullarbor dawn.

Tafoni caused by salt exhudation where ground water with salt in solution reaches the surface pores of the cave wall. The water evaporates and the salt crystallises, expanding as it does so, thus chipping off flakes of rock.

Weebubbie Cave is some 200 m long and 30 m deep. Like many Nullarbor caves, water was extracted and was originally pumped to the Eucla Roadhouse but the pipe and pump have now been removed.

Above: Cocklebiddy Cave and lake illuminated by diproton magnesium flare light. The lake is the largest underwater lake in the southern hemisphere.

Below: Two divers travelling through the 'Railway Tunnel' towards the 'Rockpile' in Weebubbie Cave.

Francis Le Guen, a French explorer, was the next diver to organise a major Cocklebiddy expedition. He travelled 20 000 kilometres from Paris to make a full-length feature film for French television. His base camp on the Nullarbor was transported by a Mercedes truck and carried half a tonne of fresh water, a tonne of food and three tonnes of equipment. The expedition used high-pressure fibreglass scuba tanks, not previously seen in Australia and manufactured by a French company. These ultra light tanks can be filled to a very high pressure of 600 bars. Diver propulsion vehicles, DPVs, capable of speeds of thirty metres per minute for up to six hours made the trip to the rockpile in half an hour. The second leg to Toad Hall took an hour and fifty minutes, with the rear diver switching off his DPV light to save battery power. The divers wore dry suits and thermolachyln underwear, so when they reached Toad Hall, they were able to sleep in their underwear beneath survival blankets for four hours. Francis Le Guen made the final dive alone, a dangerous move, which however set a new world record at 6000 metres. After a forty-seven hour dive they begin to suffer hallucinations when they returned to the party waiting at the rockpile.

The Australians led by Hugh Morrison, not to be outdone, organised another major expedition. Three divers, Ron Allan, Peter Rogers and Hugh Morrison reached a point where the tunnel narrows. It was here that Francis Le Guen had discovered the limestone skeletons of bats, which disintegrated as he touched them and a new crystalline type of flowstone, which 'shimmered like gold' as it flowed through his fingers. The three divers changed to single scuba tanks to squeeze a further 1.5 kilometres past the point reached previously. At that point Hugh Morrison took off his tank and, extending it in front of him, squeezed into a narrow tunnel for a further 240 metres. At this point, he reports, he was psychologically unable to go any further. Fear of the dark and being crushed in the bowels of the earth persuaded him to back out and begin the long swim home.

It is heartening to know that the adventurous drive which led our forefathers to this country, is alive and well. Diving offers challenges to the physical and mental capacities of men and women. Exploring new and isolated areas, under difficult conditions, is one of the few remaining challenges. As long as there are challenges to their physical and mental capabilities, men and women will face the challenge and the spirit of adventure will take us all to the unknown limits of human endurance. Such exploration is not only a personal achievement, but also a breakthrough in scientific discovery as new biological and geological phenomena are recognised. In the following chapters we will see how divers have extended our knowledge of both marine life and early history.

This is the most spectacular nudibranch we have ever found and its presence in the cool water of Botany Bay came as quite a surprise.

BIZARRE AND BEAUTIFUL

Home to a vast and varied population of animals, the sea is the birthplace for all the life on earth. Its creatures have all evolved their own characteristic features which allow them to survive in the underwater world. Every imaginable kind of colour, shape, as well as offensive and defensive weaponry is used by the myriad of marine creatures in the daily struggle for survival. Streamlining is a prevalent characteristic. A smooth outline with no protruding edges is often the norm while other slow-moving or sedentary animals of the seafloor have chosen to sacrifice speed for effective camouflage. Other features, especially among the ocean's predators, are sharp eyes, probing snouts, rapid reflexes and sheer brute power. Prey, on the other hand, rely on speed, manoeuvrability, poisonous spines, unpalatable flesh, protective disguise or particular behaviour patterns to stay alive.

Over ninety-five per cent of all the animal groups living in the sea are invertebrates or animals which lack backbones. They come in a multitude of exquisite shapes and about half of them are parasitic, minute or rather obscure animals which are usually never seen. The others include sponges, sea anemones, corals, jellyfish, worms, crabs, prawns, clams, squid, octopus, sea urchins and sea squirts. However, whereas most of the ocean's animals lack backbones, the sea's true masters are those which possess them. Fish began to appear after the invertebrates but their speed and agility quickly saw them become rulers of the surface waters of the ocean. Despite this, fish supremacy did not go unchallenged. Around fifty million years ago an ancient group of mammals were lured back to the sea by the riches it contained. These ocean-going mammals included the cetaceans or whales, dolphins and porpoises. Seals and sea-lions followed, and today these mammals stand under humans at the apex of marine life.

"I was in 35 metres of water and had been diving in a channel on a wreck and surfaced from the dive. My companions hovered at the three metre level decompression stop. Suddenly, as I looked up at them, I saw a huge form silhouetted against the surface. It was a gigantic shark and was moving slowing backwards and forwards in their bubbles. I gestured frantically at them and one of the girls looked up. She began to scream into her mouthpiece, I could hear her screams as she swam towards me and hid behind my back, holding onto my tank. It swam towards us, mouth open. It turned sideways and I saw the ridges on its back and the spots, and I knew it was not a tiger shark as I had first thought. The head sloped gradually from the dorsal fin, with the mouth at the extremity. I realised that it was a whale shark, one of the largest creatures on earth, and this was a baby, only eight metres long. The shark went out of sight and then came into sight again, scooping plankton into its huge mouth. We were nearly out of air and, as we surfaced, it came within two metres of us, splashing its dorsal fin on the surface. There was a lot of plankton in the water and the whale shark was feeding. It also seemed to enjoy playing in the divers' bubbles: the silver bubbles tickling the white belly as it swam in an undulating way through them."

'Humpback Whale and Calf' (94 × 44 cm), an oil painting by Brian Maloney, which is held in a private collection.

Whale sharks are one of the gentle giants of our waters. They may grow to twenty metres and weigh several tonnes, yet they have never attacked and have no natural enemies. They are found in the warm temperate waters of the Indian, Pacific and Atlantic Oceans, swimming just beneath the surface. Their huge two-metre mouths scoop up plankton with the help of gill rakers, and sieve out crustaceans and other tasty morsels.

This chapter is about some of the unusual and beautiful marine creatures and their bizarre lifestyles. It looks at human interaction with some of the sea's creatures and how they have fared under commercial exploitation. Some industries have developed and then waned and some are experiencing a revival based on more conservative methods of harvesting and cultivation.

On a clear, sunlit day, Kevin was fascinated to see a school of dolphins approach while he was in the water, as he reports:

> This was a rare occurrence, as they usually keep well away from divers, sensing them with sonar while out of sight and this is very frustrating for divers. On this occasion a school approached me, the leaders swimming right up and inspecting, closely moving their heads from side to side. I could feel the sonar as they whirled around me. I dived down to meet them as they cavorted around me. There were even baby dolphins. I was left with the impression that this encounter was especially for me as it was my twenty-eighth birthday. Like many reports of encounters with dolphins, I felt that I had experienced a supernatural event.

Divers often report that encounters with dolphins and whales leave them in a highly emotional state. Even sceptics report the deep feelings they experience as a result of these rare encounters. A program has been established to document these encounters. Based in New Zealand it is known as Project Interlock.

Whales, the giants of the sea, have been hunted by man almost to the point of extinction. Whaling has ceased in Australia and the way of life of our early whalers is now a part of our history. In Twofold Bay, Eden, men and killer whales worked together to hunt the migrating fin and humpback whales. Whaling began at Twofold Bay in 1828 although whaling ships took shelter in the Bay before that. It was in that year that Thomas Raine established a whaling station. Benjamin Boyd arrived in Sydney in 1842 and set up a whaling establishment, but

A watercolour from the Mitchell Library collection, of whales broaching in Torres Strait, depicted during the voyage of HMS Rattlesnake.

the peak of the whaling industry had already been reached by 1840 and the following decade saw exports of whale oil slowly decrease.

The killer whales were working with the whalers in Twofold Bay in 1876 and probably had been doing so for several years for, by then, they were individually identified by the whalers. The killers worked in three packs, each with a leader — in all there were thirty of them. When migrating whales passed the Bay, killer scouts, which had been sent out by the leaders of the packs to locate them, herded the whale pods into the bay, where the remainder of the killer whales ambushed the migrating whales until the whale boats arrived. A whale boat crew consisted of six men who chased the whale using the strength of their bodies. Once harpooned, the whale would sound and the whaleboat was drawn through the water at a furious pace by the whale. The harpooner had to strike the whale between the ribs to ensure a kill, then wait until the whale ran out of air on the bottom. Each time the whale surfaced for air and blew, the line was drawn in as the whale tired. The killers were hitting it from all directions slowing it further and sharks tore chunks from its body. Still the whale was a danger to the men in the tiny boat and they had to get past the tail and against the whale's side. Standing on the peak of the bow, the head whaler drove the needle-like killing

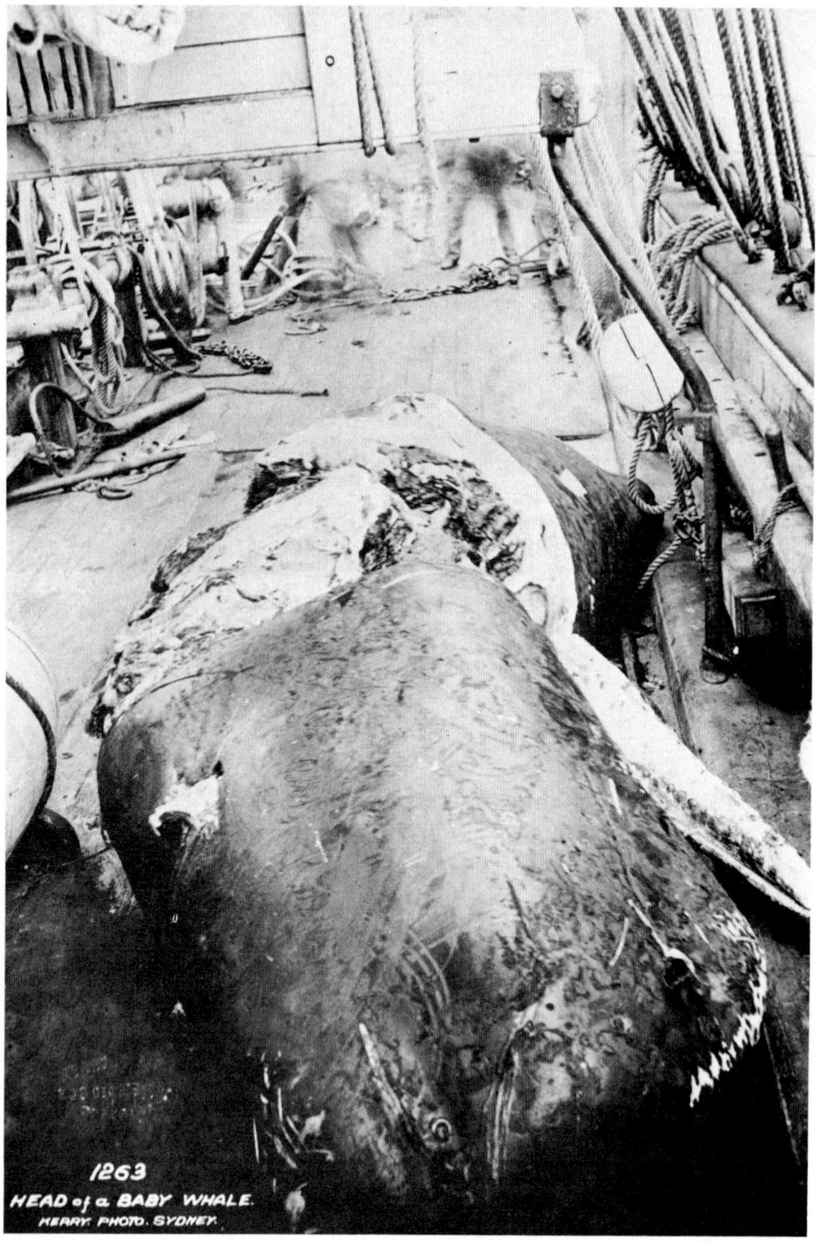

This photograph of the head of a baby whale was taken in the early part of the 19th century.

lance into the whale's body, into the heart or lungs. Again and again the boat rode up to the whale and the lance plunged into its body seeking heart, lungs or bowels, until finally it was still. The whale slipped below the surface and the killer whales received their payment — a feast on the tongue and lips, the only part of the whale they ate.

When migrating whales passed the bay at night, the pack of killers would attract the whalers' attention by loudly blowing and tail flopping. The boat crews would set out to sea, led by the killer whales. One strategy of the killer whales was to throw themselves over the blowhole of their prey, preventing it from breathing and quickly exhausting it. Old Tom was a killer whale leader. The whalemen knew him by his high dorsal with a distinctive notch. He was known to catch the line between the dead whale and the boat with his fin and tow the whaling boat around in a disconcerting way. When whaling eventually ceased in 1925, the killer whales dispersed, but Old Tom stayed in the area, until 1930, when he was found dead. He had been frail for some time. His body has been on display in the Whaling Museum at Eden since then, a reminder to all of the ability of man and whale to work together.

The southern right whale was the foundation of the Australian and Antarctic whaling industry in the 1800s. By 1900 the herds had been decimated and Australian whalers began to hunt the sperm whale in the eastern States and the humpback in Western Australia. By 1929 sightings of the southern right whale were almost non-existent and it was placed under total protection by the Australian government. A whale watch program along the east coast of Australia has been recording the migrations of humpback whales. They move north into southern Queensland waters in the last week of June and the first three weeks of July. Sightings have been increasing during the northern migration. The southern migration occurs through the same waters in the last week of September and the first three weeks of October, and by December the humpbacks are in their Antarctic feeding grounds.

Southern right whales prefer the inshore waters, which is why they were hunted so heavily and are now an endangered species. They pass Sydney in July and August *en route* to breeding grounds in the Solomon Islands and reappear in August and September on their way south, the mothers now with calves. The calves are uninhibited and will swim up to divers but the mothers are more wary. They occasionally enter harbours and bays.

More than 50 000 people turned up to see fifteen southern right whales at Warrnambool on the Victorian coast in the winter of 1982. The National Parks and Wildlife Service are conducting aerial surveys to determine how many whales migrate along our coastline and is currently formulating guidelines for divers and whale watchers, but it is likely snorkelling with

Above: Whale strandings on our coastline have been reported since the earliest days of European settlement. If the leader can be coaxed to sea, other whales will follow and the mass stranding of the pod is avoided.

Below: A humpback whale — its calf riding the mother's slipstream — is captured by photographer Dean Lee.

A reef shark is a common sight on the coral reefs of Queensland.

The grey nurse shark conjures much fear in the mind of Australians. It first got its bad reputation when an arm was disgorged from a shark in a Coogee aquarium. The grey nurse is actually fairly docile, living in rocky gutters offshore. In the early days of diving, its habit of lying on the bottom made it prey to spearfishermen, hence its numbers have rapidly decreased.

whales will be permissible.

Sharks are survivors, they have been around for 400 million years. Tiger sharks live in deep water and seldom come into the reefs where divers are, but they are the biggest sharks. The sharks' sensory system is highly evolved. They can detect low frequency sound over long distance and detect low thresholds of sound pressure. They can hear our heartbeat in the water, probably from fifty metres away. Their vision is excellent. They see colours and have good resolution, and have better night vision than we. Their sense of smell is acute. They have a unique electrical perceiving sense with which they can detect the minute electrical fields around animals and the magnetic field of the earth as they move through it, giving them a built-in compass. They can tell the whole state of physiology of an animal, from its heartbeat, and can tell if it is frightened.

In Australian waters the white pointer shark is considered the most dangerous. It was a white pointer which attacked Kevin at Currarong, distinguished by the membrane which closed over the eye as it attacked. Dangerous Reef in South Australia, has a colony of sea lions living on the reef, south east of Port Lincoln. It is a well known haunt of white pointer sharks made famous because of film sequences by Ron and Valerie Taylor which were taken on Dangerous Reef.

Seals were the basis of Australia's first export industry. As early as 1798 sealers were clubbing or stabbing seals in Bass Strait for their skins and oil. In that year, 9000 skins were collected from Cape Barren Island. In 1802 elephant seals were

The textile cone is extremely poisonous and has been responsible for several fatalities. The small fish which it preys on are susceptible to its poison. Several cone shells are venomous, the geography cone being the most dangerous.

Playful seal pups approach the diver.

plentiful on King Island in Bass Strait, but they are now found no nearer than Macquarie Island in subantarctic waters. By 1806, the supply of seals was depleted in Bass Strait and sealers were going to Kangaroo Island in South Australia and then to New Zealand. By 1857, the fur seal had been hunted almost to extinction while the elephant seal, which was valued for its oil but not its skin, remained abundant.

There are now two fur seals in Australian waters distinguishable by their size: the Australian fur seal is the largest in the world. It occurs in Victoria and Tasmania. The Victorian colonies number as many as 5000 individuals. In New South Wales seal colonies are found at Montague Island and Seal Rocks. On the west coast of Australia, the New Zealand fur seal is found, extending into South Australia at Recherche Archipelago. The Australian sea lion is found from the Abrolhos Islands to Kangaroo Island. The adult bulls have massive necks and shoulders and whitish manes, but the females are difficult to distinguish from fur seals at a distance.

Southern elephant seals breed on Macquarie Island and are rare visitors to Australian waters,

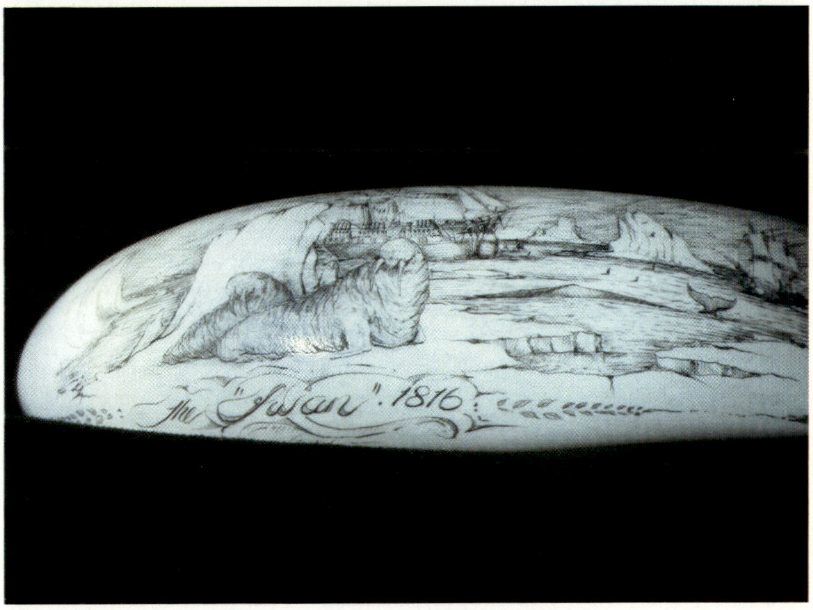

Above: The leopard seal is a rare visitor to NSW but divers found this one which had been shot by fishermen stranded at Seal Rocks. Leopard seals have a reputation of aggressiveness with divers, but zoo keepers find them friendly.

Below: Scrimshaw was the early whaler's method of recording events and places. Such art forms are now an extremely valuable record of historic events and as the taking of whales' teeth is illegal in many countries, this is a dying art. This piece was carved by one of the few modern craftsmen working in this form.

The crown-of-thorns starfish has had a population explosion in recent years but scientists believe this is a natural cycle. They attack mainly fast growing corals such as plate and staghorn and may benefit the slower growing, reef building corals by allowing more light to reach them.

but the leopard seal is a regular visitor. Young animals wander widely and often reach Australia in spring. They are dark spotted on the back and, when approached, emit a hissing or thumping sound. Leopard seals eat some quite large prey: penguins, squid seals and krill form part of their diet. The fur seals eat barracouta, squid and crayfish.

The crown-of-thorns starfish is certainly a bizarre inhabitant of our waters. Considerable controversy has raged about its population increase with headlines such as 'Threat to the Reef', 'Reef Park a Farce in Face of Starfish Threat' and 'March of the Starfish'.

The crown-of-thorns controversy hinges on whether the outbreaks in numbers have been the result of human activity or a natural event. The idea that the reef is threatened is considered erroneous by most scientists because the reef has been around for millions of years, withstanding far greater impacts such as an Ice Age without destruction. Predators of the crown-of-thorns have been recorded and certainly the numbers of the *Triton* shell, a predator of young starfish, have been depleted by shell collecting. As this is a predator of the larvae, we don't know how much impact this has on the total population of starfish. The *Hymenoptera* shrimp is another predator that has only been recorded three times on the Great Barrier Reef, but is reported as a predator of crown-of-thorns starfish in Africa. Attempts to control the numbers of starfish by stabbing them and removing them from the reef, have not had prolonged success. The crown-of-thorns was first recorded as being unusually prevalent in 1962–63 near Cairns and Green Island. Australia is not the only place where large aggregations of starfish have occurred. During 1983 the Great Barrier Reef Marine Park Authority sampled 117 reefs along the Great

Chris Deacon is wary of touching the crown-of-thorns starfish as the spines are venomous.

This large jellyfish was found at Bowen Island, Jervis Bay when a blue current brought very clear water to the Bay. Divers swim with the sea jelly observing the small fish which live in the tentacles and a species of crab which is camouflaged by its colour among the tentacles.

The Spanish dancer can be found at Lord Howe Island as well as other tropical regions. Its name is derived from the flamenco-style movements it makes, which propel it slowly through the water.

Barrier Reef. Aggregations of forty or more starfish were found on only 14 per cent of the reefs sampled. Of the reefs sampled, 86 per cent had less then forty crown-of-thorns, so it would appear that speculation that the crown-of-thorns is destroying the reef is entirely unwarranted.

The sea cucumber is a fascinating reef dweller, one that was responsible for the early exploration of our northern shores. *Bêche-de-mer* fishermen from Indonesia were visiting our northern shores before Captain Cook arrived at Botany Bay. They came in proas, a type of fishing boat, to the Kimberley region and to Cape York, where their visits were recorded by Aborigines in rock paintings. The voyage from Macassa in Sulawesi in December or January took advantage of the northwest monsoon. The Macassan camps are distinguishable by the presence of tamarind trees, which they planted for the fruit eaten to prevent scurvy. In 1978 a pilot scheme was begun to again fish for *bêche-de-mer* on the Great Barrier Reef. It was the first time the trepang had been harvested since the Macassan trade ceased in 1907. The marketed varieties are mama teata, blackfish and prickly redfish. They are eaten in Asia where they are highly valued for food and for the treatment of high blood pressure. Several other species are also edible. When the *bêche-de-mer* is disturbed, it shoots out long sticky white tangled threads, the purpose of this being to confuse or entrap the enemy. They may even break into several pieces if violently agitated, or may cast out the guts and respiratory trees, which they later regrow. The *bêche-de-mer* feed by picking up the sand passing it through their bodies like a vacuum cleaner, without absorbing it, but sifting out its living contents. They even live under the sand with just their tree-like gills poking out. Associated with the *bêche-de-mer* is the eel shaped messmate fish, or fierasfer, which emerges from the intestines of the *bêche-de-mer* at night to feed on crustacea. They live as young ones in the plankton, floating upside down. They and their hosts are certainly bizarre creatures. The beautiful colours of the body and tree-like respiratory organs of the *bêche-de-mer* are also quite attractive.

Among the beautiful creatures of our waters, the molluscs are hard to rival. The numerous brightly coloured nudibranchs are a favourite with photographers. The Spanish dancer is a nudibranch named after flamenco dancers of Spain. As it twirls its frilly edges, moving through the water, divers are fascinated. This creature is easier to capture on movie film, for its movements are even more fascinating than its brilliant red and pink colours.

Above: The sea cucumbers include the bêche-de-mer *which is fished commercially and this interesting group which are more sedentary with finely branching tentacles held upwards in the water to catch plankton. They are found on the soft sandy mud areas and on the continental shelf and are often brightly coloured. This one has hidden in the sand with just its tentacles and mouth exposed.*

Below: The bêche-de-mer *is a delicacy in the Orient and fishing for it led to the exploration of our northern waters.*

Not all molluscs are as harmless as the nudi-

This close up view of an octopus reveals the suction pads of its tentacles, the slit-like eye and the siphon behind it.

branch. The cone shell's venom is quite deadly and visitors are warned against picking up and handling any elongated shells for if it is a cone shell, the poison is quick acting and deadly. The most dangerous is the geography cone, which is common throughout the Indo Pacific region and Queensland.

Octopuses, unlike the other molluscs, haven't a bone in their bodies. They are cephalopods from the Greek for head and foot, for their feet are attached to the head and surround the mouth, an extraordinary arrangement. These solitary animals live in dens or nests which they build up with rocks and shells. From there they forage particularly at night and if they find a crayfish in a pot, it is an easy meal, for the octopus can stretch its body to squeeze through the tiniest cracks. The eight arms have suction caps on the lower surface which are operated by a fleshy piston within the sucker. It reaches out an arm and, with its suckers attaching to rocks on the bottom, pulls itself forward. When danger threatens the octopus swivels its siphon in the opposite direction and pumps out a fast jet of water. If it really wants to escape, it may release a cloud of dark, dense ink. The inky cloud may be so similar to the octopus in shape that a predator attacks the ink cloud instead of its maker. While the predator backs up in confusion, the octopus has changed into a ghostly white transparency and escapes against a sand background. The octopus ink is almost indestructible. The sepia used in colour printing once came from the European cuttlefish, another cephalopod.

The paper nautilus, a fragile and beautiful shell, is made by an octopus. These parchment-like, foam-light shapes are created as a cradle for the argonauta octopus to lay her eggs. She makes a transparent film with her paddle shaped tentacles, weaving it around her own body,

In order to move, the octopus attaches its suckers to rocks on the bottom and pulls itself forward.

which she uses as the mould to lay the clear calcareous film. It soon turns pure white and she paints the tips with ink from her ink sac. After laying her eggs, the mother carries the case in her arms or she may sit within the shell accompanied by the tiny male who is only 10mm in length.

The male octopus transfers the sperm in a sperm packet from the third right arm. He places it in the mantle cavity of the female after a courtship that lasts several hours. The female may lay up to 40 000 eggs and guards them against predators, jetting them with air from her siphon, but it is only the paper nautilus that carries her eggs around with her. Once the eggs are hatched, the beautiful cradles drift away and are broken up. Occasionally they find their way ashore unbroken. At Flinders Island in Bass Strait, the islanders patrol the beaches day and night from the end of March. Westerly winds bring them ashore and, every seventh year there is a bumper harvest collected as prize souvenirs.

One octopus is a beautiful but deadly inhabitant of our shores. It is the blue ringed octopus, a pretty creature that flashes peacock blue rings in patterns over its body and tentacles, almost inviting you to pick it up, but to do so is to risk death: the bite from her beak is almost unnoticeable yet the small dose of poison can paralyse and kill adults quickly. A toxin from the venom is used as pain relief for terminally ill cancer patients.

There are two species of blue ringed octopus in our waters and both have caused fatalities. A wide variety of crabs, fish and mammals are very sensitive to the toxin that causes muscle paralysis and a fall in blood pressure. Some reports from victims of bites show very little injury, suggesting there is a great variation in toxicity. First aid consists of application of an arterial tourniquet and artificial respiration.

Cuttlefish and squid have a reputation of

being more apt to bite than an octopus. The beak is a formidable weapon. When the cuttlefish bites, the upper mandible closes within the lower so that the sharp edges of each cut obliquely, like a pair of scissors. The beak cuts an opening in the flesh, then the tentacles are inserted to bring out morsels of flesh. The radula does not cut as in other molluscs, but assists in swallowing. The cuttle eats whiting, crabs, shrimps and prawns, including the shells. It seeks its prey in daylight catching nocturnal shrimps that are hidden in the sand by blowing with its funnel and jetting them out of the sand. Although normally a hunter, the cuttle can lure its prey with the tips of its tentacles remaining camouflaged until the last moment when its arms shoot out to hold prey, as I once found:

> Something moved in the back of a dark cave and ten arms moved slowly forward. Behind the arms were two slit-like eyes. Was the cuttlefish sleeping or carefully watching? With a rippling movement of my fingers, I enticed it to drift towards me and out into the open. It was the Australian giant cuttle which had attracted my attention. I collected some crabs and offered one to the cuttle but it remained motionless in the back of the cave. The wriggling crab struggled and got away. In a second the cuttle shot forward and grabbed the tasty morsel. Soon it was motionless again. When I let a crab out of the bag, it began to crawl towards the cuttle. The cuttle began to flash, its colours vibrating almost iridescent. It was interested. Now it would follow my hand with the struggling crab outside its cave. It emerged and began to circle me flashing reddish brown, and seaweed-like projections appeared above its eyes. It raised its upper tentacles and two long arms shot out and grabbed the crab from my fingers.

Colour change and camouflage in cuttlefish are unrivalled in the animal kingdom. The colours are under nervous control and a total transformation may take only two-thirds of a second. In many cases, movement is quick and abrupt and then the sudden flashing and play of colours aids the cuttle to pounce on its prey, or elude an enemy. The skin has three layers of

Above: Cuttlefish, like this one, generally begin mating in October.

Below: The blue ringed octopus is a deadly inhabitant of our seas, biting with its beak to inject the poison. There have been several deaths from it and its presence in rock pools and tidal areas where children play, make it likely that there will be more. Children should be warned and divers should not handle it.

coloured chromatophores, a bright yellow layer near the surface and under it, an orange red layer and beneath that, a dark base, enabling something like the play of colour in black opal. The chromatophores are small elastic bags filled with pigment, which are expanded by radial muscles and contract back naturally. When they are contracted, a pattern of fine stippling is seen on the cuttle. Expansion of all layers of chromatophores colours the cuttle like a rainbow. Embedded in the skin both below and above the chromatophores, are reflectors, called iridocytes. These minute mirrors are doubly refractive so when light falls on them, through the chromatophores, there is a play of colours enhancing the chromatophores that are controlled by messages transmitted from the eye and tentacles to the brain. The eyes are very similar in structure to our own and the arms, too, have chemical receptors and touch cells which give it a great deal of information about its environment.

Once when we were diving on an old wreck, I swam down to inspect the boiler that was covered in weed and lay on a bottom of small boulders. A movement among the boulders revealed two cuttles, well camouflaged on the almost white bottom. They were performing a strange dance, chasing each other and embracing each other's arms. It was a courtship display. It was October, allegedly the beginning of the mating season for the Australian giant cuttle. Already there were long strands of white eggs hanging inside the boiler waving in the current. Soon the baby cuttles would emerge — enchanting minature replicas of the adult, already capable of squirting ink and changing colour. Three

Above: A small cuttlefish eating a yellowtail which he has caught by shooting out his tentacles, grabbing it and biting it with its beak. At night, cuttles will catch sleeping fish.

Middle: Egg masses of the giant Australian cuttlefish are laid in caves and shipwrecks as well as in the sponge masses seen here. The giant Australian cuttlefish is seen seasonally in Jervis Bay, appearing at the beginning of winter. Large numbers of aggressive cuttles are seen at this time when they are mating. When they raise their tentacles, divers should be careful as they are large and their beaks may be dangerous. Encounters with the cuttles are a fascinating experience. In early summer the baby cuttles can be observed hatching from the eggs.

Below: The colour pattern on the giant Australian cuttlefish suggests it is curious as it approaches Colleen Lander and lets her stroke its tentacles.

Clown fish live in a symbiotic relationship with anemones, but shrimps and other crustacea are also found clinging to the tentacles or inside the mouth of the anemone.

or four other large cuttles were moving around inside the boiler. It was the first time I had seen cuttles together. I lured one into the open and it let me tickle and pat it, then it began to follow me. Brilliant iridescence would flash across its back, alternating with the light tan colour. Waving its frilly mantle with short jets from its funnel, it would swim backwards just in front of me. A violent movement would send it three or four metres away, jet propelled by the siphon. Forwards and backwards movement was accomplished by altering funnel direction. When turning, the fins would undulate in opposite directions, movement being aided by gentle jets from the funnel. For full jet streaming, the fins are folded against the underside of the body. As we swam back to the anchor rope, it followed, easily keeping up with our slow swimming. After one last pat, we watched it move towards the rocky bottom. A flash of colour and it was gone.

Sea horses are fascinating creatures that make delightful pets. They live in the sea attached to weed and are often found around jetties and piers so their numbers have probably increased with man's intervention. They prefer live food, obtained by running a net through weed to collect live shrimp, but they will also eat mosquito wrigglers and other insect larvae and can be trained to eat frozen brine shrimp, but prefer them freshly hatched. One of the amazing sights to see is the male sea horse giving birth to several hundred little seahorses in the aquarium. The babies swim rapidly about wrapping their tails around any object within their grasp, even other sea horses. They eat ravenously consuming 3000 to 4000 shrimp each per day. In sea horses the eggs are conceived by the female and during the mating process she transfers them to the male's pouch where they mature. Sea horses gradually take up the colours of their surrounds, and in the

The male pot bellied seahorse broods the eggs in a special pouch like other members of the family, the pipe fish and sea dragons. Their prehensile tail grips the seaweed on the jetties where they are often found.

This leafy sea dragon was found hovering above the weeds at the anchor line at the end of the dive. Sea dragons blend effectively with the kelp beds making them difficult to find and easily overlooked. They may be more common than divers believe for this reason. They are not fast swimmers and are easily approached and handled.

There are two forms of sea dragon in Australia. This one, which is very weedy in appearance, occurs in SA, the Gulf, WA and is sometimes found in Victoria. The more northern species is the leafy sea dragon which is found on the NSW coast, and is more colourful with orange, blue and brown markings on its body, though it does not have the surreal splendour or the distinctive body shape of the South Australian one. As with sea horses, the eggs are carried along the underside of the tail by the male.

In the deep sponge gardens around Jervis Bay hundreds of swimming anemones attached to sea whips wave in the current like a field of wheat in the wind. They are filter feeding from their vantage point on the sea whips above the gardens.

aquarium should be provided with something to cling to.

The leafy sea dragon and pipe fish belong to the same family and all have an armoured skin and prehensile tail with which to grip the seaweed on which they live. They are slow swimmers and fun to watch, bizarre denizens of our southern waters.

Anemone fish live in association with large tropical sea anemones. In subtropical waters they clothe the bottom in places. Equipped with specialised stinging cells, or nematocysts, the anemone's tentacles can paralyse fish that come too close. Yet the anemone fish living amongst the tentacles remains unharmed, and even relies on the anemone for protection from predators. How is the fish immune to the anemone's stinging cells?

The young anemone fish, after living for a few weeks in the plankton, seeks out a suitable anemone. Most anemones are already saturated with occupants who fiercely repel the intruder. When it finds an unoccupied anemone, the fish must acclimatise to the deadly tentacles. At first, it makes only brief contact with the anemone, but after a day it can safely remain in the tentacles. A chemical change in the mucous coat of the fish renders immunity, by gradually slowing the firing of the anemone's nematocysts. There are twenty-six species in the anemone fish genus and eleven of them are found on Australian coral reefs. Some are associated with particular anemones. The pink anemone fish is associated with Ritter's anemone, which usually has an adult pair and three or four smaller fish living amongst its tentacles. The adult fishes constantly harass the younger ones, establishing a pecking order. Many anemone fish begin their lives as males and later transform to female. The female is usually larger in size and spawns throughout the year at monthly intervals. The eggs are deposited in a small nest at the base of the anemone and the male is very active in taking care of the nest, fanning the eggs with his pectoral fins. The eggs hatch on the night of the seventh day of incubation, the larvae struggle to the surface where they are swept out to sea. In a few weeks they begin their life-long association with an anemone.

Many fish change sex during their life: wrasses, parrot fish, angel fish, gobies, damsel fish, bream, snapper and emperors, mostly change from females into males. In the tropical moon

Above: This little transparent fish was captured on film by photographer Jim Dyer.

Below: A colourful swimming flatworm undulates through the water.

wrasse there are two types of males: those that are born as males and those that have changed sex. There is a balance of reproductive success between the two types of males because they have different success rates on reefs of different size.

On small reefs, brightly coloured blue moon wrasses with long pectoral fins dominate all smaller fish. These may be males or females but are drab green and all the smaller fish are

The multicoloured gills of the Christmas tree worms extend from the coral in tropical waters.

submissive to the big male. The wrasses spawn on the diurnal high tide every day in summer on these reefs. About an hour before spawning, the blue colour becomes permanent in the dominant male and he begins swimming up to the surface and down again: he is looping. Then he begins a rapid quivering of the body and tail whilst fluttering his bright pectoral fins. This behaviour eventually attracts a female who rises off the bottom and the two fish rush together towards the surface, release eggs and sperm and rush back to the bottom, while the eggs are swept out to sea. If the dominant male is killed or lost, the next largest fish rapidly changes to a blue colour and begins attacking females. If this fish happens to be a female, after ten to twelve days, she becomes aggressive and blue and if caught and examined can be shown to be degenerating her ovaries and developing and producing sperm. She has functionally changed sex, and is now a dominant male.

On larger reefs with extremely high densities of moon wrasses, group spawning is the norm. All the wrasses move to the one spawning site which is usually in the lee of a projection, out of the current. The smaller males that don't stand a chance on the small reefs, gang up against the dominant male. There may be twenty or thirty of them and they form together in a tight school around the spawning site to watch the dominant male display to the females. When a female moves towards a blue male to spawn, the other males join in and the whole group release their gametes in a white cloud. Many females choose to spawn with the groups of males and do not bother with the big dominant males. Individuals on large reefs do not improve their reproductive success by becoming dominant males and one wonders why they bother to change sex at all on large reefs.

There is a balance between the success of the two types of males on reefs of different sizes. The dominant males have an advantage on small reefs where they pair-spawn with all the females and prevent any other males from reproducing. On the larger reefs with group spawning, the smaller males have an advantage so both types of males are able to survive within the species. This phenomenon in moon wrasses was discovered by Gillian Eckert, and it will be interesting to see if it applies to other sex-changing fish.

Cleaner stations on the reef are sites where fish go to have the parasites removed from their bodies. The coral-banded shrimp or a blue and black wrasse can be seen minutely examining a large coral trout or cod that waits patiently. The fish opens its gills for inspection and has its parasites removed.

There are nine species of banded coral shrimps in Australia. The most common is called banded coral shrimp and its red and white striped nippers and long white feelers are often visible at cave entrances or overhanging ledges, hanging off the ceiling. Shrimps are very territorial but will tolerate a mate. The real cleaner shrimps belong to another family and many individuals share a ledge and work together in the cleaning station. The red and white striped *Lysmata amboinensis* has no common name. It occurs throughout the tropical Indo Pacific region and in eastern Australia as far south as Montague Island, preferring deeper water than the banded coral shrimp. Unlike the latter, the true cleaner will leave its ledge and walk all over its client pecking away.

Bramble cay is Australia's northern most piece of land, a sandbank in Torres Strait that is a nesting site of the green turtle. It is only one of many turtle nesting sites on Australia's Great Barrier Reef. There is even one site on the mainland, at Turtle Sands, Bundaberg, but that is way south of Bramble Cay. It is inhabited by seabirds, and a unique species of native rat — the Bramble Cay mosaic tailed rat, which has the most restricted distribution of any Australian mammal. Applied Ecology Pty Ltd has been investigating the nesting of green turtles on Bramble Cay to determine if it is a source of eggs for a turtle farming venture. Waves from storms cause erosion of sections of the sand spit and the loss of an estimated 90 000 turtle eggs in the wet season, a consistent seasonal loss over a period of years which is natural wastage. Turtle harvesting is an integral part of the culture of the Torres Strait people and harvesting is restricted to local consumption among island and coastal communities. The harvesting of nests, laid in areas of the beach in danger of erosion, should not affect the natural production of hatchlings. The cay is on the trading route between the Torres Strait island of Daru and Port Moresby, New Guinea. Turtles are harvested by ships' crews illegally. As the turtles are female, their harvest removes reproducing individuals from the turtle population. Such harvesting has destroyed many turtle rookeries throughout the

world. Perhaps turtle farming would supply a need without this wanton destruction.

The green turtle comes ashore on coral sands at night and when she is well above the tide's reach, begins to dig into the sand with sweeps of her flippers. As she digs, sand flies and her flippers slap against her shell. A pit three or four metres wide and half a metre deep is created, with her in it. By now she has dug down to the damp sand in which she begins to excavate an egg chamber. Her hind flippers act as scoops as she digs a chamber fifty centimetres deep: her egg chamber. She folds her hind flippers over the hole as her soft ovipositor begins to release eggs coated in mucous with anti-fungal properties. She lays about 100 eggs, then covers the egg chamber with wet sand and begins infilling the pit with her fore flippers, obscuring the nest site. Then she slowly drags herself back to the sea. She stays in the area feeding until another clutch of eggs is ready, in about two weeks. She may nest ten times during the nesting season from

Above: The green turtle is found on the Great Barrier Reef where it can be seen during the day in the water and at night on the coral cays where it comes up the sand beaches to lay its eggs.

Below: A close up portrait of the green turtle as it carefully watches the photographer.

Above: Another species of butterfly cod with short spines which are webbed to make a fan of exquisite colours. This is a small fish which may be an immature cod.

Below: The transparent sea squirts have their body structures visible including the pharyngeal basket which filters a large volume of water, extracting the plankton on which it feeds. The colony is highly specialised, the individual animals being high on the evolutionary scale and close relatives of fish.

October to March.

The eggs incubate in the sand from fifty to sixty days, then the first turtle hatches, which stimulates adjacent eggs to hatch. The hatchlings struggle to the surface remaining just below the surface until it cools, early in the evening. They emerge from the sand and speed down the beach with whirling flippers. The crabs and birds attack them on the beach and sharks and fish attack once they are in the water. Thus the turtle eggs are an important food source for many animals on the reef. Night herons breed to coincide with the period of greatest turtle hatching. Thus it is possible for 1000 pairs of night herons to breed on another turtle rookery at Raine Island on the northern Great Barrier Reef and to raise three or four chicks on the baby turtles and eggs. When the turtles leave the island at the end of their breeding season, the herons, too, disappear. But for every hatchling turtle eaten, hundreds survive. The simultaneous hatching and emergence ensures there will be some turtles who survive the predators, both those on land and in the sea.

The plankton or floating life of the sea is the basis of the food chains with the plants — diatoms and other algae — the major energy producers in the ecosystem. Algae are not immediately evident on a coral reef, but are present both in the tissues of the coral and enmeshed in the skeleton of living corals. Direct transfer of organic food between the algae in its tissues and the coral, enables the coral to be less dependent on plankton in the water. Corals on reefs in the open ocean depend on algal photosynthesis within their tissues for food more than those in richer coastal waters where zooplankton are more numerous. In very deep waters, the basis of the food chain may be bacteria able to use the sulphur emitted by undersea volcanoes. The prolific life around such undersea sources has only recently been discovered in the deep ocean. The floating life of the sea includes not only the microscopic inhabitants but also the larvae of many species and even adult animals that spend all their lives exploiting this rich food source.

Bluebottles or Portuguese man-of-war are colonies of individuals. A gas-filled sac acts as a float for the colony and may be up to thirty millimetres wide. The feeding is performed by the nutritive polyps, each of which has only one long tentacle that may extend several metres down from the surface. The tentacles are armed with stinging cells which, as anyone who has encountered one knows, can be painful. The stinging cells trap surface fish such as mackerel and flying fish that are paralysed and drawn to the mouths of the feeding polyps, where digestion is shared by all members of the colony. Such bizarre co-operation among individuals of a colony is not limited to bluebottles.

The stone fish occurs in Queensland and at Moreton Island in very large numbers. It lurks in dark places, camouflaged from passing fish which it gulps down if they approach too close. Although poisonous it is not a danger to divers but it is to reef walkers who are likely to inadvertently step on it. The dorsal spines are like hyperdermics, injecting venom into their victim.

Salps are much more specialised colonial animals related to ascidians. Most live in tropical and subtropical waters on the plankton and may be anything from a few centimetres to metres in length. The individuals are oriented in the wall of the colony so that one opening is to the outside and the other empties into a central canal. Circular muscle bands produce contractions of the body wall which drive water through the colony giving it control over its movement. Divers have reportedly swum through such jelly-like colonies that may be up to ten metres long.

Living on the plankton and often associated with blue currents is the sunfish that lives on bluebottles, jellyfish and other drifters. Its dorsal and anal fins act as propellers but sometimes, the fish lies on its side as if basking in the sun. They may have a girth of over eight metres and the increase in size from larvae to adult is probably the greatest in the animal kingdom.

Among the many venomous Australian fishes are catfishes, stingrays, stone fishes, scorpion fishes, rabbit fish and the butterfly cod. The striped catfish forms spherical shoals up to a metre in diameter. Only some of the spines, on the first dorsal fin and the leading edge of each of the paired pectoral fins, has the barbs that inflict the venom. If a hand is waved through the school, the catfish will most likely part rapidly and not touch the diver, but if he should handle one, and the spine enter the skin, the venom is released into the wound. Pain is instantaneous and may last for from thirty minutes to more than two days. The swelling may be massive, involving the entire limb and accompanied by primary shock.

Even more poisonous, but less likely to affect

The mosaic moray eel is unique to the Lord Howe Island area and is one of the most spectacular species in Australia.

A magnificent butterfly cod which is about 35 cm long and has its venomous spines erect. They associate with schools of small silver fish on the reef.

divers, is the stonefish. It resides on the bottom, camouflaged among the mud or rubble and in coral reef pools where a direct landing on it could penetrate a wetsuit or flipper easily. However its habitat is one where divers are less likely to be than reef walkers or fishermen who have been the main victims. Depending on the depth of entry of the spine, a point is reached where the venom gland's contents are expelled with great force into the puncture. Since this depth is one centimetre for fish twenty-six centimetres in length and several spines may inflict the venom, the results can be excruciating. The degree of pain indicates the severity of the sting. First aid involves application of a pressure bandage to retard absorption of the venom, immersion in hot water which de-natures the protein, injection of emetine by a medical practitioner or in its absence, dousing with congo red, methyl violet, dilute hydrochloric or acetic acid or potassium permanganate. An antivenene is available and two millilitres is capable of neutralising the effects of up to two stings.

Another fish that is less toxic, but more likely to affect divers, is the butterfly cod. The intense pulsating pain may radiate within three to ten minutes to the extremities, the wound become pale and its edges discoloured. The severe pain lasts three to eight hours during which nausea, primary shock, increased perspiration and diarrhoea may be evident. Swelling of the wound and enlarged lymph glands may persist for days.

The bullrout and fortescue are small fishes which may also give a nasty sting if accidentally handled, and even rabbit fishes can cause severe pain.

Whilst fish can be deadly, the greatest threat in our waters is the box jellyfish or sea wasp. A tropical species of jellyfish, it occurs as far south as Gladstone and has resulted in the sudden death of fifty or more Australians to date. It appears in shallow coastal waters in late November each year and disappears by late April. Even during the summer season when it is inshore, the population may disappear for days during periods of south easterly winds. It is thought that they move into deeper water and descend to the sea floor when conditions are unfavourable in shallow waters. Death can occur within minutes of being stung. Pain increases during the first fifteen minutes in mounting waves despite removal of the tentacle. First aid has in the past included the application of tourniquets, to prevent the venom from entering the general circulation. More recent first aid techniques prefer binding the wound. Alternatively dousing areas of contact with aluminium sulphate or a solution of Stingose, or with dilute acetic acid are helpful. External cardiac massage and artificial respiration may be needed and medical attention should be sought immediately, to administer the antivenene. The best protection is to wear clothing so that the trunk, arms and legs are adequately covered, then the waters can be entered at any time with impunity.

Moray eels are much maligned fish that look far more ferocious than they really are. They readily accept food from divers and soon recognise those who feed them. Well known underwater photographer, Valerie Taylor, became very fond of Harry, a moray on Heron Island.

> At first he was very nervous and used to pretend he was a very ferocious dragon — opening his mouth wide and waving his head around. But I persuaded him to stop that nonsense and after a while he became so docile I could reach into his hole and pull him out, and have him twine around my arms and neck and peer into my mask.

Morays are not normally aggressive and will take food from the diver's hand. There are many varieties of morays in Australian waters, and most are fish eaters hunting at night by the scent of their prey. Because they lack a swim bladder

Moray eels are well known cave dwellers, usually found in small caves. They look more ferocious than they actually are.

Exploring the Yongala *wreck, one encounters an enormous variety of marine life. The big stingray lives on the wreck and investigates the divers who enter his domain.*

Catfish school for protection and are found in masses like this from the tropics to southern waters. They have poisonous spines and though it is tempting to wave an arm through them to see how they close around the diver without touching him, it should not be done without wearing protective clothing on the arm and hand. Their whiskered faces peering at the camera make an entertaining sight.

The saltwater crocodile lives in the estuaries and mangroves of northern Australia. The number of crocodiles has been reduced by hunting but although they are not endangered they are now protected in Australia.

and are bottom dwellers, retiring into the coral by day, divers rarely appreciate how many different species there are. On Heron Island reef, twenty-three species have been recorded. There are other species in cooler waters such as the green eel of Sydney that owes its colour to minute plants in its slime. If you are cleaning abalone on the rocks this moray will come right along small crevices with barely enough water for it to move. Related to the morays are the pike eels which are distinguishable by the presence of pectoral fins and the long canine teeth, spoon shaped snout, and yellow eye. Snake eels are another bizarre eel burrowing into the vents and bodies of fish or writhing in clusters in the water. They are sometimes mistaken for sea snakes.

Sea snakes are, of course, air breathers and have scales, which the eels lack. There are five species found in southern Barrier Reef waters. There are pelagic species that are washed up after storms, and others that live on coral reefs or on sandy and muddy inshore areas.

Crocodiles used to be very common in the Queensland and northern Australian tropical mangroves. They were even found on the offshore reefs and islands where they would come ashore on the atolls when moving along the coast. The saltwater crocodile is greatly feared by the Aborigines and the people of Papua New Guinea. They have been known to watch people coming to a waterhole, and to wait for them. If a person came to the same place over several days, the crocodile might be waiting right at the spot and attack, dragging its victims underwater where they drown. They can also move at speed across land and it may be difficult to outrun them in some country. Once they have taken human flesh they may attack again. Livestock is lost to crocodiles when coming to water to drink. The crocodiles grab the beast by the snout and hold it underwater to drown. Then they put the carcass under a rock or log until it rots, before eating it.

Crocodile hunting was a popular activity and the skins were quite valuable. Some incredible characters and stories have emerged from the folk lore of the 1940s. While Aboriginal people hunted crocodiles and feasted on the meat and eggs, it was not until 1945 that commercial hunting began. Hunters went out at night with spotlights and shot the animals for their skins. Over 87 000 were traded until 1958. In 1964 hunters were licensed and in 1971, after twenty-six years of commercial harvesting, saltwater crocodiles were protected, although their protected status is presently being reappraised.

Originally the saltwater crocodile was widely distributed in Asia, from India to the Philippines and along the south coast of China. Its range is much more restricted today, the result of hunting for skins and meat and because they were considered vermin.

In recent years they have been considered a public threat in Darwin Harbour, so the decision was made to catch them and relocate them on crocodile farms. The Edward River Crocodile Farm in Queensland is now marketing crocodile meat to restaurants and delicatessens. The meat tastes like chicken or veal. As well as meat marketing which is a new venture, crocodiles are harvested for their skins. As there are not tanneries in Australia able to tan the leather to the standard required, they are exported overseas.

In Australia, the saltwater crocodile inhabits tidal coastal rivers and swamps which contain fresh water. The areas they occupy have distinct wet and dry seasons, and the monsoon brings the greatest mortality, to the eggs and young crocodiles. They breed in the freshwater swamps and billabongs, which are a popular recreation area. The nests are in tall grasses beside permanent fresh water and the females prefer to nest in close proximity to where they are at other times of the year. The sex of the crocodiles is determined by the incubation temperature, or rather, the rate at which the embryos develop. The young crocodiles move from tidal mainstream to freshwater but the movement of adults is little known.

The pearling industry in northern Australia is another industry which has seen remarkable recovery. When the pearling industry began in 1869, pearl shells were common in waters only a metre deep. In the 1870s pearl fever gripped northern waters. The shells were collected and loose pearls removed, then left in the sun until the remaining pearls could be easily washed out — up to seventy-five from each shell. Pacific Islanders were kidnapped for both diving and plantation work. In 1875 an area of 8000 square kilometres was being dived for pearl shells, employing over 700 native divers at ten shillings a month. As the shallow waters were depleted, diving suits were introduced and the boats worked deeper waters further from port. By 1880 the easily accessible Torres Strait beds were worked out, and in the 1890s, several pearl shelling areas were closed to allow the oysters to regenerate. In the race to get more shell, many divers died either from the bends, or as a result of coming up too quickly, and divers were often

These baby crocodiles huddle together in the protection of a crocodile farm in Australia's north.

buried in their heavy canvas diving suits. In 1893 with the discovery of rich pearl beds in deep water off Darnley Island, many divers died on the 'Divers' Graveyard', many of them were Japanese divers. Torres Strait's peak pearl shell production came in 1897 when there were 260 boats employing 1600 men and Japanese and kanaka divers were earning sixty pounds a month. The pearl shell industry died with the introduction of plastic buttons, but it is appropriate that the Japanese are now producing cultured pearls of world renown in our northern waters. A cultured pearl oyster operation on the Coburg Peninsula, Darwin, employs ex-abalone divers to harvest pearl shells for a cultured pearl farm at Port Essington. There are five cultured pearl farms in Australia, including one near Thursday Island in Torres Strait, and they harvest shell from northern waters to grow pearls by artificially seeding the pearl shells under cultivation. Some reports of high mortality in the West Australian pearl oysters are being investigated to determine the degree of threat to the industry.

Australia has many beautiful and bizarre underwater inhabitants. Divers over the last two decades have increased our knowledge of the oceans enormously, yet there are still discoveries to be made. The sea has been described as the last frontier on earth. We have seen how commercial exploitation can reduce the population of marine animals such as whales and seals to very low and uncommercial levels. Other ventures in pearling and crocodile farming, when managed on biological principles, can be financially viable industries.

The fate of many ships is to wind up wrecked on a coral reef. The family who owned this yacht were wrecked only 40 km from their destination after sailing from Europe. Sailing at night when close to their point of arrival, they went up on the reef and the yacht quickly became a total wreck.

SHIPWRECKS AND SKELETONS

The story of early Australia seems to be a history sprinkled with a liberal number of shipwrecks. But only a very small number of these have proved to be the remains of the treasure ships of legend. Over 300 years ago, the Dutch East Indiaman *Gilt Dragon* foundered on a reef north of Perth on her way to Batavia. Her rich haul included not only an immense array of practical items; she was also reputed to be carrying eight chests of silver, worth almost eighty guilders each. Other, more mundane vessels — traders and transport ships — have subsequently shared the fate of the *Gilt Dragon*. One, with a more sinister purpose, was the slave trader *James Matthews*, which came to grief at Cockburn Sound in 1841.

Over the centuries vessels from Spain, Portugal and Britain, as well as from Holland, were lost along Australia's coastline. Today, these old wrecks are slowly dissolving time capsules of a distant past and important lodging houses for large communities of fish. Australia protects its historic wrecks with an Historic Shipwrecks Act. This Act prohibits diving on any of the declared historic wrecks which dot the continent's coast without specific permission of the relevant state government authorities.

As well as all the old ships lying broken on the ocean bottom, Australia has a considerable number of relatively modern corpses. In the 1960s various state government fishery authorities added to these by using tyres, derelict cars and old ships to create artificial reefs. The reason for this was that fisheries biologists found that, within six months of sinking a ship, sufficient varieties of fish were attracted to effect pronounced increases in fish catches by amateur and commercial fishermen.

"When I first plunged into the water at the site of the Dunbar *wreck, I was surprised that anything had survived the century of pounding by the ocean waves. It was some time before I recognised that the bottom did look slightly different in the area of the wreck. It was so obscured by time and marine growth encrusting every surface, that it was only the unnatural straight lines that made me realise I was on the wreck site. Then I began to notice other man-made objects, the straight edges of the ballast blocks and the anchor. I was diving thirty metres from the rocks and it was as close as I dared go in the swell that was crashing and thundering against the cliffs. Later I was persuaded to explore the base of the cliff face, among the ledges and crannies where small items from the* Dunbar *had washed into the crevices. The bottom was cemented into a solid mass, and it was only by chipping away at it that items such as coins or jewellery were found."*

The Dunbar *was wrecked very close to its destination when the captain missed the entrance to Sydney Harbour and ran into the cliff face at the Gap, in bad weather. There was only one survivor who scrambled ashore from the bowsprit. The other passengers were washed onto the inhospitable cliffs and drowned.*

Australia's isolation from the roots of her colonisation in England, led to an early dependence on shipping that is hard to appreciate today. When the First Fleet arrived in Sydney, they encountered a foreign land completely alien to their previous experiences. Food was abundant yet they starved through lack of experience in cultivating or harvesting it: an idea as alien to us, as their experiences must have seemed to them. They were so dependent on ships arriving that the loss of every ship was felt by the early settlers as a personal one. Even when they did not have friends or acquaintances on board, they felt the loss in deprivations that resulted from losing the cargo carried on the ships. Of course the dependence on shipping meant that they could also envisage their own danger, should they visit friends and family back home in England. While the loss of an airliner today is a tragic event, we do not feel the loss in as personal a way as our ancestors did when a ship was wrecked sometimes within their sight and almost at its destination.

The immigrant ship the *Dunbar* was on her second voyage under the command of Captain Green in 1857. She carried a crew of fifty-nine with her sixty-three passengers, most of whom had relatives in Sydney, her port of destination. Her cargo consisted of coins minted for the developing colony of New South Wales, established in 1788. The exact number of gold sovereigns and copper and silver coins in her cargo has never been determined, but it is estimated at between 10 000 and 48 000.

Eighty-two days out from Plymouth, England, during the night of Thursday, 20 August 1857, disaster struck. Gale force south easterly winds had been blowing for most of the week, causing mountainous seas along the coast. At about 7 pm she passed Botany Bay, about ten kilometres south of her destination and was estimated to be about fifteen kilometres out to sea. She was close hauled on a starboard tack heading north north east in an east south east gale. Coming further up the coast, South Head light could be seen between squalls. The ship made several points of leeway, her true course being north north west and her speed three knots. When South Head light was abaft the beam, the *Dunbar* was reckoned to be about ten kilometres outside the entrance to the port. In fact, she was no more than four kilometres from land. The following account is from the Aust. Encyclopaedia.

The Dunbar *anchor was salvaged by local residents in 1907 and built into the cliff face at the Gap in Sydney's Watsons Bay as a memorial to the 121 people who lost their lives in this maritime tragedy.*

> All hands were called on deck and the ship was kept away to the westward for the entrance, running before the wind with her yards almost square and South Head light slightly on her port bow. The foresail was hauled up and a very sharp lookout kept for North Head which was not seen; a few minutes later however, breakers were seen right ahead. At once the master knew he was close in to Outer South Head, ordered the helm to be put hard aport and the yards braced sharp up on the starboard tack, in hopes of clawing off the land again to the north-eastward, and his orders were promptly carried out. But, with wind and sea setting her strongly to leeward, the ship was too close in and carried too little sail, and, though she started to turn to the northwards, about two minutes later (shortly after midnight) her port bow struck the rocks and she was hurled almost broadside onto the cliffs just north of the signal station, midway between the lighthouse and the Gap.
>
> With the first shock the topmasts went overboard. Directly afterwards the sea broke over her starboard side, carrying away people, boats, bulwark and masts.

Within a matter of minutes the once-proud ship, an outstanding vessel of her day, had become a mass of wreckage while those on board had been swept to a sudden and awful death in a maelstrom of pounding waves.

The Dunbar, *an immigrant sailing ship went aground on Sydney's doorstep in 1857 with loss of all but one passenger. The intrinsic items which have been found scattered under ledges by divers exploring the wreck over a number of years include: sovereigns, gold jewellery, personal effects and the ship's equipment which has also been salvaged.*

At first light the next day, bodies and wreckage were being washed up at the Gap and along Sydney Harbour foreshores. It was soon apparent that all must have perished. Mrs Graham, the wife of the lighthouse signal master, had a strange dream on the night of the tragedy. When the storm was at its height, she awakened suddenly and called to her husband, 'Go down, Jim, and rescue the poor fellow in the sea.' Mr Graham thought that the fury of the storm had unnerved his wife and soothed her to sleep, but once again she awoke and entreated the rescue of the man at the base of the cliffs. Naturally, on such a night it was impossible for the lighthouse staff to do anything. The spray from the waves flew high above the light and their pounding shook the building. The most remarkable part of Mrs Graham's dream was yet to come, for when the sailor James Johnson was found clinging to the cliffs thirty-six hours later, she recognised him instantly as the man she had seen in her dream, struggling wildly in the surf.

The rescue was carried out after he was seen by Mr Spinks of the White Hart Inn. A rope was lowered to the sailor, who was pulled to safety in an exhausted state. Able Seaman James Johnson was the only survivor of the 122 people on board the *Dunbar* on that fateful voyage. He never went to sea again, but joined the lighthouse service and, while stationed at Newcastle lighthouse, aided the rescue of the sole survivor of the *Cawarra*, a steamship that sank in 1866 on Oyster Bank.

There may have been an error in judgment in the ship being so close to shore at night in such bad weather, but Captain Green and his officers were not blamed for the loss of the *Dunbar*. He had not mistaken the Gap for the entrance to

Bow and torpedo tubes of a deep submarine off Port Phillip Heads, Melbourne.

Above: Coins from various countries have been found on the Dunbar, *some with a hole for hanging them on a watch chain, though this Bolivian coin was apparently in use as currency.*

Below: A display of Dunbar *relics photographed in the early days of diving on the wreck. The square porthole frames a brass keg tap with gold plated spoons, watches, dentures, rings, lockets and the numerous silver and gold coins in the foreground.*

Sydney Harbour as some had said, but might have run into port because he knew from the leeway the ship was making that he could not possibly weather the sea, and would inevitably be blown ashore and wrecked within a few hours. Hence, his only chance for safety was to run into port and anchor. Had he continued only one mile farther north before rounding for the entrance, he would have taken his ship safely in, and this slight error in reckoning was most probably due to the current setting to the south south west.

The wreck was discovered by divers in 1955, when divers of the New South Wales Explorers Club saw the pig iron ballast which marks the site. Off the bow lay two big anchors, welded by time into a mass of twisted chain and boulders. Camouflaged cannons appeared as two-metre cigar shaped forms. The wreck was a congealed mass of wood, pottery, cutlery and all the other items carried by a 19th century time capsule that the ship had become. Lying on the rocks at the base of the cliffs, it is only in good weather that it is possible to dive on the wreck. Huge swells over the site in six metres of water predominate throughout most of the year. On early winter mornings after a still night, conditions are ideal. Divers sought relics such as coins cemented into the bottom with chips of porcelain. Minted for the young colony, they bear the date of the voyage, 1857. The pennies and halfpennies were not minted by the Royal Mint as might be expected, but by a business firm: the Australian Tea Mart. They were known as 'tradesmen's tokens'. In Australia and the United Kingdom, merchants were encouraged to mint them to overcome a shortage of small change that occurred between the 1840s and 1860s. They were legal tender and there are 700 known types. The *Dunbar* tokens bear the symbol 'Peace & Plenty'. Their value has soared since the wreck when they were worth only a penny. Coins from all over the world have been discovered on the wreck. Gold sovereigns were highly prized. Other gold items recovered included a set of dentures, complete with spring attachment to the jaw, gold plated spoons, lockets, rings and engraved watches. Gold tarnishes very little underwater and is readily cleaned. Divers began blasting the site in an attempt to recover more gold. Everything else shatters under the impact leaving piles of broken rubble.

In May 1966, the South Pacific Divers Club raised one of the anchors. It was a two-metre sheet anchor that was carried amidships. The two large bow anchors lay twenty metres north. They were raised using forty-four gallon drums that, in the swell, were difficult to handle and dangerous. When filled with air the drums lifted the anchor to the surface where it was guided away from the cliffs and towed to Camp Cove near South Head. The anchor is part of a memorial in Beverly Hills, Sydney, while a cannon was displayed in Marineland, Manly.

Mugs, gold and silver cutlery, spoons delicately embossed with fruit, kitchen serving ladles and cake knives with a large 'D' engraved on the handles: all the items carried by a passenger ship of the last century, are among the relics recovered. Taps in brass, once plugged into wooden kegs used in the officers' quarters. Embossed buckles and buttons from the crews' uniforms and lockets and watches engraved by relatives are reminders of the human drama that unfolded. Among the relics is a pair of small silver spurs, that must have been for a little boy to use in his new homeland. One wonders what part the compass with spirit level intact, might have played in the tragedy. Shotgun butts, bullet moulds, lead bullets, cannon balls, a trajectory sight for a cannon were other items recovered.

Most amazing was the discovery of human bones. This is very unusual on a wreck, but divers found the skeleton embedded in pitch which had apparently flowed from a broken cask over the body of a youth aged between fifteen and twenty years as the ship sank. A report of the find states:

> We have been uncovering leg and arm bones from time to time, but our most amazing find occurred when we created a cavity in some pitch; peering inside, my partner and I could see a complete skeleton. Leaving it almost undisturbed we left, intending to return and photograph the find at the next possible opportunity.

The finders of the relics were obliged to deliver it to the Receiver of Wrecks and to hope that the government was generous.

Allan Robinson, the discoverer of many of the earliest Dutch shipwrecks in Western Australia, describes his experiences in a book called *In Australia Treasure is not for the Finder*. It is an exciting book, that shows how one man's determination and incredible drive led to the discovery of wrecks dating from before European settlement and describes his legal battle to claim ownership of objects recovered. At that time, the generally accepted date of first discovery of Australia was Captain Cook's voyage. Allan Robinson changed all that. We now know about the Dutch ships that visited our coast in the 17th century; few know that there are possibly ships of even older dates wrecked on our coastline.

The figurehead of an unknown vessel which met its fate on the reefs of north Queensland. The figurehead of a woman is carved from a single piece of timber and was washed onto a remote beach where it remained intact, untouched by shipworm and destruction.

The mysteries of the Mahogany ship and of the Chinese wrecks have not been resolved.

Mulvaney in *The Prehistory of Australia* states his belief that the Chinese visited our shores. When Flinders visited Arnhem Land in the *Investigator* in 1803 he met a fleet of proas. The commander, Pobassoo, told him there were over 1000 men working the Australian waters that season. Flinders met the Macassan trepang or *bêche-de-mer* hunters at the Malay Roads. The fishermen sailed in wooden proas of ten to twenty-five tonnes out of the Celebes harbour of Macassa, catching the strong trade winds that carried them to Australia in about fifteen days. The Macassan proas were equipped with oblong fibre matting sails, bamboo slatted superstructures and two steering oars. Records of their visits were recorded by the Aborigines, who also began to wear headbands and smoke a Malayan-type pipe. At Anuru Bay an archaeological site has been investigated where the proas landed and boiled *bêche-de-mer* in cauldrons then dried and smoked them. A reported Chinese wreck on Long Island, Queensland, has never been adequately investigated.

Legislation in 1964 gave little protection to the archaeological importance of shipwrecks. The Receiver of Wrecks was responsible for locating the owner of the wreck and if not found in twelve months, he had to dispose of the wreck at auction. There was no provision for compensation to the finder of the wreck and the Receiver had no obligation to apply emergency conservation to objects recovered. Tracing the legal owners of wrecks from 1875 or earlier was a difficult and time consuming task. The law was inadequate for effective operation once the advent of scuba had made its impact.

The government of the Netherlands held title to wrecked vessels of Dutch origin under their own law and could make agreements with divers, a provision which was not in existence in Australian law. In 1972 the Netherlands transferred its right to Australia which agreed not to seek costs of searching or recovery from the Netherlands, whilst accepting their continued interest in objects recovered.

The Commonwealth Historic Shipwrecks Act 1976 is subject to State proclamation that the Act applies in waters adjacent to the coast of that State. While most States have now declared that the Act applies, New South Wales, has not done so. Australia is one of the first countries to assert jurisdiction over historic shipwrecks on the continental shelf. The validity of this in international law is doubtful, although it may be a trend that will be followed by other countries.

The Act requires persons who find the remains of a ship, or articles associated with a ship, to give notice describing the article and its situation to the Minister. The Minister may grant a permit for the exploration or recovery of the historic shipwreck or relics. The disposal or damaging of the relics or their removal outside Australia is in contravention of the Act and punishable by a fine of $5000 or five years imprisonment. The Act requires persons with historic relics in their possession to give notice describing the article and its situation to the Minister.

Divers who notify the Minister of the location of any remains of a ship may receive a reward, but the article must first be declared an historic shipwreck or relic, and the notice must be the first to lead to a particular wreck or relic. The amount of compensation is determined by agreement between the Commonwealth and the person from whom the property is acquired. If no agreement is reached, that person may take action against the Commonwealth in the High Court or the Supreme Court of the State.

The implementation of the state *Historic Shipwreck Act* in Western Australia meant that the Act was proclaimed for waters adjacent to Western Australia in 1977 and thirty-eight historic shipwrecks declared. The Japanese submarine I–124, located off Darwin, was declared an historic shipwreck because it was viewed by both the Japanese and Australian governments as a war grave. The *Aboriginal and Historic Relics Preservation Act,* 1965 has provided some protection to historic shipwrecks in South Australia, but new legislation is in preparation. The *National Parks and Wildlife Act* 1970 was used by the Tasmanian government in 1977 to protect the wreck of the *Sydney Cove*, by declaring a State reserve over an area between Rum and Preservation Islands in Bass Strait where the wreck lies. Queensland declared the Act to apply to the waters adjacent to its coast in 1977 and protected the *Pandora* site on the day it had been discovered. The movement to protect cultural heritage on the seabed has now received worldwide attention, with Australia leading the way, and the case *Robinson* v. *Western Australian Museum*, a trial case, documented in Allan Robinson's book, was the impetus to all this.

Allan Robinson salvaged a large quantity of cargo from the wreck of the *Gilt Dragon* in 1963. The West Australian government passed legislation in 1964 and this gave the wreck to the West Australian Museum. Robinson took his case to the High Court and won the legal battle, but his triumph was shortlived when the Commonwealth passed the *Historic Shipwrecks Act* 1976. With his untimely death in 1983, many divers believe details of wrecks he discovered have been lost forever.

The anchor from explorer Matthew Flinders's ship was found by divers in Encounter Bay (which is now known as Victor Harbour) south of Adelaide, and is on display in the South Australian museum.

From the time of European settlement on the east coast, ships have been wrecked, remaining on our shores as time capsules. In an historic sense this was far preferable to them going home to England to become hulks on the Thames or broken up for salvage. We still have the *HMS Sirius* in which some of our convicts were transported on the First Fleet. The *Sirius* lies on the seafloor at Point Ross, Norfolk Island where she was wrecked on 19 March 1790.

The *Pandora* was carrying fourteen of the *Bounty* mutineers in leg irons when she went down in 1791. Four of the prisoners went down with her when officers refused to remove their manacles. The remaining survivors sailed to Timor in the ship's boats, a long voyage of 2000 kilometres, even today. The logbook of the Commander, Captain Edward Edwards has also come to light. Apparently it remained in the possession of the Edwards' descendants until it was given to the Admiralty Library. The wreck was reportedly found by Dr Colin Roderick in 1960, when he snagged his anchor on the bottom and swam down with an aqualung to release it. More recently in 1977, Ben Cropp's nephew Ron Bell and Steven Domm reported finding the wreck and received the $10 000 reward. It was the first time it had been seen in 200 years and was declared an historic shipwreck on the day it was found. The federal government passed responsibility for the administration of the Act to the Queensland Museum and a $60 000 grant and private sponsorship enabled expeditions to explore the site. She lies in thirty-three metres of water near Pandora Cay, north Queensland, in an area patrolled by bronze whaler and tiger sharks. During the first expedition, 686 dives were made by the excavation team and during the second, in 1984, another 800. The most exciting find was a silver watch possibly belonging to George Hamilton, the surgeon, whose diary has provided much information on the voyage. His apothecary items of mortar, ivory syringe and tourniquet clamp were recovered as well as twenty-four cannons and a chain pump wheel, thought to

Objects become an integral part of the ocean floor as the encrusting corals solidify it into the bottom. This photograph shows a conglomerate of artefacts compressed and solidified into the seafloor, retaining the basic shape of the chain within the mass.

have been the faulty item that led to the *Pandora*'s sinking. The expedition involved considerable logistic problems in keeping divers to strict decompression limits. This was aided by the use of a $15 000 computer sponsored by NEC. The computer was also used to establish, from grid co-ordinates, exactly where items were found. An exhibition of the *Pandora* relics will coincide with Bicentenary celebrations in Australia.

Matthew Flinders was shipwrecked on the Great Barrier Reef. After charting most of the coast, he was on his way home with his charts on the merchantman *Porpoise*, when she ran onto an uncharted reef at night. The *Cato* which was accompanying them went onto the reef but the *Bridgewater* managed to escape. Those survivors of the two wrecked vessels landed on a small sand cay where, in the morning, they hoisted a white ensign upside down as a distress signal. Though Captain Palmer of the *Bridgewater* must have seen the inverted flag, on arrival at Bombay he reported all hands lost. One voice disputed his story. Mr Williams claimed that he heard Flinders shout 'For God's sake Captain Palmer do not desert us.' He was left behind in Bombay, which was his reward for his honesty, as the *Bridgewater* sailed off never to be seen again. Flinders was able to repair one of the ship's boats, a cutter, which he sailed to Port Jackson with twelve men averaging 100 kilometres a day with sail and oar. Flinders returned to Wreck Island (off Rockhampton) in the *HMS Cumberland* six weeks after he had left and rescued those on the island who had been under the strict command of his brother, Samuel Flinders.

Immigrants to South Australia from Ireland were nearly at the end of their journey after being cooped up in the *Nashwauk* for eighty-nine days when they broke into the rum casks. The crew may also have been imbibing because they mistook the lights of Dalkeith, the home of a local hotelier, for the lighthouse at Semaphore. When they ran aground they tried in the early dawn light to refloat the ship, but the vessel soon had three metres of water in her hold and the

Silver pieces of eight from the Gilt Dragon *have been made into jewellery.*

Wreck divers Ken Wendon and Peter Gubb descend the anchor chain to a broken section of the sailing ship Centurion. *Very little remains of most such old wrecks, though the* Centurion *is more intact than most as it lies in 20 m of water in Sydney Harbour.*

passengers were brought ashore or walked. Bad weather inevitably followed. The ship became a total wreck and the sands of St Vincent Gulf covered her. About eighty years after the wreck, a Mr Charles Robinson at Moana, in Adelaide, used six horses to recover the anchor. It now graces the caravan park at Moana. The sand that covered the wrecksite preserved the wooden vessel and her cargo of china plates and hand-made bricks for the masonic lodge in Adelaide. When the sand moved, as it does at regular intervals after winter storms on our coasts, the wreck was exposed and divers from the Underwater Explorers' Club were able to bring items of the cargo ashore. Cups and saucers inscribed 1854, a year before the ship sank, were perfectly preserved after over 100 years.

The passengers on the *Admella* which went aground in 1859 on Carpenter's Reef were not so fortunate. The ship lay across a cleft in the reef, visible from shore. Rarely have the survivors of a shipwreck been so tortured by vain hopes. The boats were swept away but a young Dane saw a ship's lifeboat caught in wreckage and tried to swim to it with a rope. He made unceasing efforts to reach the ship but was swept to sea and drowned. A sister ship passed to the westward but failed to realise there were people stranded on the rocks. They said the people looked like seals. Many died of starvation and exposure while attempts were made to rescue them in the rough sea. The wrecksite today reveals little but the copper ingots scattered across the rocky reef.

The nature of the site of the wreck is important in determining its historic value. Where the bottom is sandy, as we saw with the *Nashwauk*, valuable and delicate items were found in good condition. If the wreck founders in sheltered waters, preservation will be better. One of the oldest intact shipwrecks in the world is the *Wasa*, which sank in Stockholm Harbour where she was on her maiden voyage in 1628. She was raised intact in 1961, her timbers preserved by the mud and lack of boring *Toredo* worms in the cool northern waters.

The barque *Centurion* left Sydney Harbour in the morning of 17 January 1887. The fastest of the Aberdeen White Star Clippers, she did the passage to Sydney from England in sixty-nine days on her maiden voyage. Her owners expected she would beat the fastest clipper in the world, the *Thermopylae*, but she never did. As she grew older her prestige sank from wool clipper to general cargo and her rigging was altered to the more easily handled barque style. Thus, when she left Sydney on 17 January she was carrying coal. With a strong easterly gale blowing, another ship had been forced to anchor between the Heads and was drifting close to North Head. The *Centurion* unable to make way into the strong wind, was in tow of the steamer, *Phoebe*. The *Phoebe* cleared the anchored vessel but her tow the *Centurion* was carried towards it, and hit the rocks inside North Head. She slid off the rocks into eighteen fathoms of water. When discovered earlier this year, she was lying in twenty metres on a mud bottom. She is a composite ship built of timber on iron frames and copper sheathed to protect her against Toredo worm which would have slowed her speed. Wreck researcher, John Riley, established her as the wool clipper *Centurion* ninety-eight years after her loss in 1887.

The Sydney coastline is not the only area dotted with shipwrecks. Melbourne's Port Phillip Bay was equally notorious. In fact there were so many that a bell stood at the Heads with a sign 'only to be used in the event of shipwreck'. The greatest shipping tragedy in Victoria's early days was not in Melbourne but along the stretch of coast now called the Great Ocean Road, on

Above: The Centurion *was one of the Aberdeen clipper ships which sailed the England to Australia run with such famous ships as the* Cutty Sark. *She was blown ashore and sank in 18–20 m of water off North Head and was not discovered until 1985 by Sydney divers. She provides an interesting wreck site with iron frames and fittings strewn about the bottom.*

Below: The Centurion *being an Aberdeen clipper ship, carried sails on masts which comprised several sections. In this mast section the lower portion may have been the end stepped into the hull.*

As the wooden wreck sinks to the bottom, the wooden parts are eaten away by toredo worms and only the iron frame and hardwood parts remain unless the sand covering them protects them from boring animals.

the south west coast. On the night of 31 May 1878, the iron clipper *Loch Ard* went down.

She was running before the wind and when the reef was sighted she was too close to it. In a final bid to avoid the coast, her anchors were dropped. All efforts to save the ship failed and she struck Mutton Bird Island broadside. A lifeboat was manned and one of the survivors, Midshipman, Tom Pearce, was trapped beneath it when it capsized. He held on underneath until it righted itself. When the boat drifted towards shore, a wave washed him off and he managed to swim ashore. He had been washed into the narrow entrance of Loch Ard Gorge and even more amazingly, two hours later, he heard a cry for help and saw a young woman clinging to a spar. He swam to her rescue and struggled to shore where they collapsed. Later he climbed the cliffs and reached help. Cargo and bodies washed up in the coves and cliffs.

A Geelong syndicate purchased the remains of the wreck and cargo at auction. A large amount of the cargo disappeared under cover of darkness, but in the gorge, the syndicate had guards to protect their property which mounted eight-foot high along the shore. Another storm washed much of this back into the sea, but a sale was held in the customs shed at Geelong and further goods were sold as they were salvaged. Among the items salvaged was a ceramic peacock produced by the Minto Potteries in Staffordshire England, a fine art piece on its way to the first International Exhibition in 1880. When divers found the *Loch Ard* in 1967, such delicate pieces had been pounded by heavy seas since 1878 and there was little hope of their recovery. When Stan McPhee found the wrecksite it was the end of a five-year search. He found the bow, then on the other side of the reef lay the hull. News of the discovery leaked out and

Above: The National Parks and Wildlife Service office in Port Campbell camping area has a display of Loch Ard *relics and this anchor is displayed outside the office.*

Middle: The cargo carried by the Loch Ard *included items for the immigrants in the new country. The pewter teapot in the centre can be contrasted with the one on the right, which is how it looked when the diver discovered it on the bottom, encrusted in marine growth.*

Below: A mass of musket shot cemented into the bottom as it was recovered from the Loch Ard *wrecksite. The spoons and forks were part of the cargo headed for Australia.*

The remote Loch Ard Gorge was where the two survivors of the Loch Ard *disaster were washed ashore.*

divers were soon raiding the site. Legal action against two of them brought an end to the mass scavenging but not before considerable amounts had been salvaged. Today, the National Parks and Wildlife Office at Port Campbell has an interesting display of items recovered more recently and one of the anchors rests outside.

The waters of Torres Strait were dangerous hazards, particularly as ships carried only a magnetic compass. Until the end of the 19th century, there were only two shore lights, three light vessels and thirty-four unlit beacons at Booby and Goods Islands. The *Quetta*, a steamship which was sail assisted, of 2254 tonnes, was bound from Australia to England when she struck the uncharted rock. Today she is a declared historic shipwreck.

The *Quetta* has held a fascination for me since I was a child. I lived next door to a daughter of a Torres Strait Island pilot Arthur Dabelle. His daughter, now very elderly, sleeps in a bed made from the wheelhouse of the *Quetta*. The house has now been classified by the National Trust as it is one of the very early homes in Mosman. Apparently, the wheelhouse had been sitting on Thursday Island for some time when it was rescued by Arthur Dabelle. During my research for this book, I established that the *Quetta* is indeed missing the wheelhouse. She lies virtually intact and you can see where she was ripped open like a can of sardines by Quetta Rock, an uncharted coral covered rock in Albany Passage, the inner route through Torres Strait at the tip of Cape York Peninsula, Queensland. Divers can reach the wreck only at slack water or neap tide as a three-to-four-knot current runs over the wreck at other times. She lies on a swept coral bottom with sandy patches. It is an interesting dive, though the water visibility is best at twenty metres and then only from January to February. There are some intrinsically interesting items, bottles and jars, brass lights and portholes. The same currents, which hamper divers today, caused the terrible loss of life on the wreck. The *Quetta* sank within a few ghastly minutes of striking the rock. Survivors clung to anything that would keep them afloat as they were washed backwards and forwards in Adolphus Channel. Next morning seventy survivors reached the mainland where residents of Somerset found them and sent word to Thursday Island by telegraph. Of the 292 people on board, 133 lives were lost. Only three of them were women. One of the survivors was a baby girl whose parents could not be found, so she was adopted by Captain Edmonson Brown of the Torres Strait Pilot Service and she lived with him as Quetta Brown on Thursday Island. Another woman who survived describes the night:

> The sea was calm and the moon was shining. A vibration ran through the ship like a shiver, accompanied by a dull, grating noise like the muffled tearing of iron. Just as suddenly the ship began to list and there was an acrid smell like a fire smells when it has been doused with water.
>
> The ship began to settle down by her bows, and step by step my mother and I climbed up the slanting deck towards the stern. My mother kept praying, 'God help us, God help us' and I remember that I kept my arm about her until suddenly the sea rose up and engulfed us. I seemed to sink for ever — deeper and deeper — and when I came to the surface it was only to be pressed down again by the struggling black bodies of scores of Javanese passengers. Finally I broke free and began to search for my mother, I never found her. Both my parents were gone.

The *SS Yongala* went down without a trace in 1911 and it is only now that we know she sank further south, off Cape Bowling Green. She began work on the Sydney-Fremantle run carrying cargo and passengers during the gold boom in Western Australia. In 1907 she was transferred to the Melbourne-Cairns run maintaining a coastal service with her sister ship and others of the Adelaide Steamship company. She left Mackay carrying 667 tonnes of cargo and 120 passengers and was sighted as she passed Dent Island but was never seen again. A cyclone was raging in the area and the lighthouse keeper at Dent Island saw her pass during what he described as mountainous seas and a force seven to nine gale. The search continued for several days as news of wreckage filtered in. In 1943 and 1947, navy vessels detected the wreck. The wreck is only fifteen metres below the surface and alive with spectacular marine life. It seems to pulse with life as swarms of baitfish move in clouds around the divers. Groupers live in natural caves below the bow and stern, the largest weighing an estimated 320 to 450 kilograms. Divers have been feeding them with the result that as soon as one enters the area, one grouper nicknamed Baggy barges straight for the diver carrying the food and tries to swallow the bag whole. Large bullnose stingrays hover above while coral trout trevally and angel fish come in

Middleton Reef is in an area of strong ocean currents near Lord Howe Island. As it is only just above tide level it has claimed many vessels. Freighters and Japanese long liners, or steel fishing vessels have often come to grief. Some are almost intact, abandoned after the crews were rescued. Although the very exposed location makes them soon disintegrate.

On the wreck of the Foam *are many ceramic bracelets which played a part in obtaining native slave labour. They were traded for slaves and now lie cemented in the coral bottom but Chris has discovered one lying loose and is trying it on.*

to feed. Turtles resting on the deck, when shaken by the diver, give an exhilarating ride.

Blackbirding or the kidnapping of South Sea Islanders for the sugar plantations of Queensland was operating from about 1847 to 1904. The first natives, or kanakas, were from Tanna in the New Hebrides. They were brought to Australia by Benjamin Boyd for his sheep stations on the Monaro in southern New South Wales. Other natives were recruited for cotton plantations and pearl diving. Technically, the natives volunteered to come to Australia to work in the cane fields for three years at three pounds a year. When their time was served they would be taken back to the their islands loaded with trade goods bought with their wages. The fee advertised in Brisbane of seven pound per head for the best and most serviceable natives was paid by the planter, to the recruiter to cover passage money. As the demand for the native labour increased, reports of kidnapping reached official circles. The potential labourers were lured aboard recruiting vessels by a display of trade goods, bundled below deck and the hatches slammed on them. Canoes were run down and the occupants dragged struggling from the water. In 1868 the Queensland government passed the *Polynesian Labourers Act* but this was not enforceable outside Queensland and blackbirding continued. In 1891 there were still 9362 kanakas in Queensland out of a total of 46 387 introduced since the beginning of the practice. Not all the kanakas were unfairly treated, some were paid their contract wages by plantation owners and returned to their islands. Others chose to remain in Australia.

The *Foam* was a wooden sailing vessel engaged in the blackbirding traffic. She left Townsville on a recruiting mission in the 1890s but never reached her destination, for she ran up on Myrmidon Reef out of Townsville. She was washed over the reef into the lagoon making salvage impossible. The crew constructed a raft and sailed back to Townsville escaping without loss of life. The fact that she settled in the lagoon and broke up, has kept the ship reasonably intact in a confined area. When sports divers located the *Foam*, it was declared an historic shipwreck site and its history probed by the Maritime Archaeology Unit of the Queensland Museum. Among the unique artefacts are the ceramic armbands that were traded with the natives, and clay pipes that were valued after the European introduction of tobacco. The wreck is in six metres of water in a calm lagoon, beautiful coral and giant clams make the site an attractive area for sports divers to explore.

Many coastal trading vessels that carried produce to port in the days before rail and road freight, were lost on our coasts. The rivers they entered are now often so silted up, that even much smaller vessels have trouble entering. These busy little ships made such an impression on the people they served, that their loss was sadly felt. The *SS Casino* traded between Melbourne and Port Fairy for almost fifty years. Ten of her crew were drowned within 300 yards of the beach almost within reach of the Apollo Bay jetty where she was to tie up. After cables from the vessel fell short of the jetty, Captain Middleton let the anchor go and tried to swing her stern around so that he could run a line to the jetty, but the *Casino* would not swing in the swell. He backed out but struck solid bottom. He tried to beach her but she grounded only 270 yards from the beach. A lifeboat was launched but overturned in the heavy sea. Another lifeboat was overturned and finally a seaplane was contracted to fly over and drop a line. But the vessel rolled over and the last of the five clinging to a rail were wrenched away by the waves before the plane came over the hills above the beach. Originally a three-masted topsail schooner, she was converted to steam, the sails assisting to keep down fuel consumption. She made 2500 trips on the Port Fairy run and the people of Apollo Bay and Port Fairy have erected memorials, a reminder of her valued contribution to the development of their communities.

The Illawarra and South Coast Company was established in 1903 out of pastoral companies shipping interests. They concentrated on the south coast of New South Wales trade. Among them was the *John Penn*, a sail paddle wheel steamer found in 1982 in Broulee Bay, and declared an historic shipwreck under the Commonwealth *Historic Shipwrecks Act*, even though New South Wales has not accepted the legislation yet. The *Merimbula* was built for the southern parts of the run, Bermagui, Tathra and Eden yet she went aground at Beecroft Head, Point Perpendicular, where her boiler rested on the rocks while she slid into deeper water.

The coastal town of Newcastle, founded on lime burning and coal, was linked to Sydney by sea. The discovery of gold in California led to a

Above: The anchor of the SS Casino *is displayed where she sank in Apollo Bay which was a regular destination on the coastal trading route she plyed, along the Victorian coast.*

Below: The SS Casino *was a coastal trading vessel which supplied coastal towns with essentials and took their produce to market, long before rail and road transport took over the coastal trade. As a measure of their esteem for those who lost their lives while plying the coast, the people of Port Fairy erected this memorial with the ship's propellor and bell.*

The Yongala *is one of the best wreck dives in Australia. Black coral grows on the hull and schools of tiny fish shelter among the coral and wreckage as Chris Deacon descends from the surface.*

Divers will often find coral encrusted relics when exploring wrecks.

demand for Newcastle coal. In 1865 more than 870 ships from the Pacific coastline put into Newcastle for coal and reached a peak of 1425 ships in the 1890s. Convicts had built a breakwater behind Nobby's headland and the ships were lined up three and four deep to load coal. This was a booming age of shipwrecks too. More than 200 ships were wrecked on the short strip of coast north and south of Newcastle, twenty-six of them on Oyster Bank, four of them on one night. The French windjammer wrecked in 1904 was built into the other breakwall where it remains today.

The *Wandra* was a timber twin screw steamer which was sail assisted when she struck a huge swell which made her unstable. She was a small ship built on the Manning River in 1907 and carried timber to the mills, a cargo she carried on her last trip, in 1915. She was able to continue towards Point Perpendicular at Jervis Bay where she dropped anchor off Drum and Drumsticks. The crew rowed ashore as the *Wandra* sank and they landed on a small beach. They were able to climb up to the Point Perpendicular lighthouse. The *Wandra* is a popular dive today. About 200 metres off Drum and Drumsticks drop off, her bow is against the reef, and her stern out on the sand.

Sixty-milers or coastal steamers plied the New South Wales coast. With the decline of sail they began serving the coal trade from the Hunter River and the south coast. One of them, the *SS Tuggerah*, often called the Marley Wreck, rests off Cape Three Points. Another is the *Birchgrove Park*, off Avalon. They both lie in forty-five metres of water on sand bottom, a haven for fish in the desert around them.

Gold has an allure which has resulted in feverish activity since the 1850s in Australia. More recently the advent of metal detectors has added new impetus to land based prospectors, and divers have recovered gold from streams and rivers using dredges.

The *Catterthun* was a steamer engaged in the China trade when she left Sydney on the afternoon of 7 August 1895. She was carrying among her passengers many Chinese returning home. In a raging south easterly gale and poor visibility she struck a rock off Seal Rocks, New South Wales. She shuddered but continued steaming for about twenty minutes. The direction in which she steamed has caused controversy amongst divers ever since. Less than two weeks later, a salvage team, in the charge of Captain John Hall of the Underwriters Association, left Sydney. The insurers were keen to recover the £10 000 in gold that she carried. The two divers on board were Tom Briggs and William May. They found the rock, one of many in the dangerous area, and began sweeps with a wire rope across the fairly even bottom. They soon realised the vessel had steamed for longer than had been reported and widened the search area. They discovered she must have been steaming for three quarters of an hour before she sank, for they found the wreck in thirty to thirty-one fathoms. Diver Briggs made three descents the next day. It was like looking at a ship through a thick mist, he said. The vessel was lying north and south with a slight list to starboard. He felt the gold could be recovered but would need diving dress of the very best quality from Heincke of London. In due course this arrived with the summer, the very time when the north east winds accelerate the current that runs along the coast. They decided to wait until the following winter. At that time, the deepest salvage operations carried out had been the recovery of £90 000 from the *Alphonso XII*, in 26.6 fathoms. The depth to the vault beneath the chartroom where the gold was stored on the *Catterthun* was 27.5 fathoms. To determine whether the divers could work at that depth, the underwriters put them to a practical test. They were taken outside Sydney Heads, lowered to the twenty-six fathom level, and required to each fill a small bag with sand from the bottom, a feat which was easily performed.

They set out on 4 May to salvage the gold, and found the buoys they had left previously. Diver Briggs reached the chartroom and could distinguish the hatch in the floor that led to the gold. By then the ship had already begun to break up — the topmasts were gone, the rigging rotted away, the funnel lying on the deck and part of the bridge missing. The weather changed and, as provisions were low, they returned to Sydney. Leaving Sydney again on the 30 May they were dogged by bad weather. A diver did get down once, but the current on his air pipe and line made it impossible to maintain his footing on the deck. On the third attempt they set out on 29 June. The buoys had been carried away and they made 2000 soundings before they found the wreck. They still did better than the modern divers who have been searching since

Diver Rick Latimer explores the bow section of the Catterthun. *A case of gold sovereigns still lies buried in the collapsed hull.*

the 1950s to relocate the wreck. They attached a wire rope to the telegraph and the divers set a charge of dynamite to the lock on the trap door, but the detonators failed to act and a fresh supply had to be obtained from Sydney. They salvaged the bridge binnacle and compass, the compass from the chart room, and the log reel as souvenirs. The charge had only opened the trap door about ten centimetres. Bad weather again dogged operations for twenty-nine days. On 7 August Briggs inspected the chamber with an electric light encased in a stout glass cover and saw the boxes of gold — ten of them on a shelf, a metre from the man-hole. They were so close to their depth limit that had the boxes been in the bottom of the hold they may not have been able to recover them. Weather again caused problems and diver Briggs was carried off the deck landing on the sandy ocean floor in 30.5 fathoms, a record dive in Australian waters at that time. On 17 August, the current was slack and the divers put in the hardest day's work they had ever done. From 11 am till 5 pm they were never out of their diving dresses. In all they made. nine descents. The man-hole was not big enough for the divers to enter, so they got an augur 7.5 feet in length and, with this, groped among the enormous mass of paper padding that had been used to pack the treasure boxes. When they felt a box, they screwed the augur into it and brought it up into the chartroom.

The first two were taken out by May in one dive. Briggs next secured a box and May, going down again, added one more to those already in the chartroom. He tried again to get hold of the big box containing £5000 but failed. Briggs then turned his attention to this box. Five times he got hold of it, but each time it broke away with its own weight.

The next day May had not recovered from the previous day's exertions but Briggs made two descents, bringing up another three boxes. The total amount recovered was now £7350. Another two boxes were recovered by Briggs in the following days, making seven of the eight boxes of sovereigns accounted for. No attempt was made to recover the £2000 of gold taken on board by one of the Chinese passengers, or the gold nuggets in boxes. Diver May was still suffering 'the effects of the pressure' and unable to go down, so the salvors weighed anchor and returned to Sydney. The amount recovered was £7944, of which the two divers received £1191 12s between them at the rate of fifteen per cent. The total cost of recovery was about forty-five per cent of the whole.

Although some knowledge of caisson disease, or the bends, was available at the time, these divers obviously did not know about it. One wonders what happened to May, who was suffering the effects of a bad bend.

Many efforts by Sydney divers searching for the *Catterthun* have been unsuccessful. Even equipped with echo sounders, she has proved elusive. We attempted to locate her in the early 1970s and found a ship shaped reef that came up on the echo sounder, like a wreck. Diver Barry Steele found the wreck on the first sounding on the first day of search, after months of planning and library research in 1978. He reported: 'It was light enough to see fifty feet in front, the hull was still intact but the bridge and superstructure had collapsed. They recovered full bottles of wine beer and soda. The wine was pretty awful, but the beer still had a head on it.'

At current reckoning, the remaining gold sovereigns are worth $150 000. This is definitely not a dive for the inexperienced.

The first Australian ship to go down in World War II was the Burns Philp passenger cargo ship *Macdhui*. In June 1942, while she was discharging cargo in Port Moresby harbour, Japanese aircraft attacked Port Moresby for the sixty-first time. *Macdhui* became a prime target. She got underway and zig-zagged around the harbour, but received a direct hit which caused several casualties and substantial damage. She struggled back to the main wharf and evacuated but the next day was attacked again and this time she drifted onto a small reef and keeled over where she remains today. Australians are not generally aware of the number of ships sunk in our waters during World War II. Three ships sunk off the Australian coast were offered for auction by Geoff K. Gray Pty Ltd. The ships were sold to an Australian businessman by the US government in 1956. Two of the ships, the 2595 tonne *Florence D* and the 3200 tonne *Don Isidro* were sunk in enemy action off Bathurst Island near Darwin in 1942. They carried ammunition and Douglas aeroplane parts. The 7061 tonne *William Dawes* was sunk by torpedoes in 1942 off Merimbula. The cargo included copper wire and lead ingots as well as army equipment said to be worth more than $2 million.

The Australian corvette *Warrnambool* struck a mine inside the Great Barrier Reef in 1947 while engaged in minesweeping activities after the war. She lies off Cape Grenville about fifteen kilometres off the coast in about thirty metres of water. Kevin initially dived on the *Warrnambool* to salvage scrap metal. Visibility of ten metres was limited but a wealth of marine life exists with sharks, fish, sea snakes and other animals making an exciting dive.

Sydney Harbour was attacked by midget submarines in 1942 and mines were laid by the Germans as early as 1940 outside Sydney Harbour and at the approaches to Melbourne. These caused a number of shipping fatalities, and then Japanese submarine patrols captured the *Iron Chieftain* between Sydney and Newcastle, the *Iron Crown*, the *William Dawes* mentioned above and many others. Diver Jim Lester found a wrecked submarine off Disaster Bay near Green Cape on the New South Wales south coast in 1967. Another lies in sand at Seal Rocks Beach, while Melbourne has several in Port Phillip Bay.

The J class submarines built in 1915 were once the fastest in the world and were built to counter German U-boat activity. Six of them were presented to Australia after World War I. They had a surface speed of nine knots and a submerged speed of 9.5 knots. The submarines were operational until 1922 when they were put into reserve. As was the fate of many war ships they were sold for scrap. The Melbourne Salvage Company stripped the J1, J2 and J4 submarines of valuables and scuttled them in 1926 and 1927. Two others were placed as breakwaters on the shores of Port Phillip where they can be seen today. The scuttled submarines are now a diver's site off Barwon Heads, where they lie in deep water. The last submarine to be rediscovered was the J4 that was found in 1984 by the Southern Cross Dive Club. When diving at the

Above: The Malabar, *a shipwreck which gave the Sydney suburb its name, went aground during the Depression.*
Below: The coastal steamer Merimbula, *went aground on the rocks at the northern end of Point Perpendicular.*

The old ferry Sydney Queen, *was in tow when she broke away to be wrecked on the beach at South West Rocks, on the north coast of NSW.*

Diver explores the interior of a deep submarine off Port Phillip Heads, Melbourne.

Some of the wrecks at Moreton Island are quite intact and divers can explore the decks and enter the hulks which are adorned with soft coral and inhabited by fish.

deep submarines, divers may experience a cold thermal layer about six metres above the bottom. The brilliant sponges and the stunning yellow zooanthids on the conning tower make this a fantastic dive. Fish, too, are plentiful with beardies, ling, cod, butterfly perch, bullseye and roughies swarming through crevices and sergeant baker and dusky morwong in the kelp on the upper deck. Another submarine, in contrast to the J4, has less fish but it is possible to swim inside the hold, and there are weed fish on the red algae growing around the conning tower.

Shipwrecks, while not as common in these days of radio direction finders and satellite navigation, do still occur. The *Cherry Venture* went aground at Rainbow Beach, Qld and despite attempts by salvors to free her, is still there. Another wreck, the *Sygna*, went aground in 1974 at Newcastle's Stockton Beach where salvors refloated her. She broke in two as Japanese salvage expert K. Yamada had anticipated. The bow was moored in deep water and the stern remained on a sandbar. After the stern was refloated, the tug unions insisted on a flow-on of the high rates paid to the unions working on the beach salvage. While negotiation proceeded a storm smashed the stern beyond repair. Kevin was on an expedition to *Sygna* to determine the feasibility of moving divers to remote sites quickly, by using helicopters, and to perfect operations and streamline procedures in event of a real emergency. Two helicopters with several divers flew from Sydney, photographed the wreck, dived and returned. The results of the exercise established the feasibility of this method. The vessel lies just off the beach in a deteriorated state. The hull is badly rusted and split but the helicopters could land on the deck. Divers were able to gear up, stand on the floats, be carried across the water and let down in a suitable area. The helicopter cruised four metres above the water surface and the divers dropped into the sea. To return they were lifted directly from the water. The helicopter could tow a diver, rafts or salvage equipment on top of the water or suspended underneath. The exercise established that divers using helicopters are a practical combination.

There is another type of ship being sunk nowdays: hulks that have reached the end of their useful life as sea going vessels are being used to create artificial reefs.

The Japanese were the first to realise that fish are attracted to old wrecks, and to exploit this by creating artificial fish habitats. They began in 1775 with the fishermen of Manzai village in the Province of Awaji, south of Kobe, sinking large wooden frames mounted with sandbags, intertwined with bamboo and wooden sticks, into sixty metres of water on rather barren fishing ground. They were so successful that in the following ten years they sank several hundred such shelters. In Australia, it was not until the 1960s that the idea of creating artificial fish habitats was investigated. It was found that old car tyres were very suited to reef construction. Derelict cars and old ships with tyres and concrete rubble were used to create a reef at Tangalooma inside Moreton Island, Brisbane. Another nearby at Cowan Cowan consists of the tug *Melbourne*, sunk in 1979, and eight other vessels sunk to form an artificial reef. The wrecks have attracted corals, fish and many animals seeking shelter or a surface to grow on, and naturally many large predators patrol the area. The sea floor consists of sand dunes of fine white sediment.

Another large ship reef has been constructed off Long Reef, Sydney, beginning in 1976 with the sinking of the old Manly ferry, *Dee Why*. Several other vessels have joined her on the bottom and fishermen report good catches of snapper, mulloway, teraglin and kingfish. The

Aerial shot from a helicopter, of the half section of the wreck Sygna *which drifted back onshore after an unsuccessful salvage attempt.*

The Runic *was driven high and dry onto part of the Middleton Reef where she remains today. Middleton Reef lies in a remote area north of Lord Howe Island.*

Cinematographer Jim Burr photographs the marine life which now inhabits the old Sydney Harbour ferry Dee Why, *on the artificial ship reef at Long Reef.*

main purpose of artificial reefs is to attract fish to otherwise barren marine areas. Amateur anglers particularly benefit from the increased productivity but there are enormous economic benefits to Australian seaside communities, in the increased fishing and tourist potential. Even six months after placement, there is a great variety of fish present on the reefs and as the marine community develops, so does the fishing potential.

Divers benefit by the provision of interesting dive sites and old ships, whether sunk accidentally or by design,as they are fascinating places to dive.

Divers have established new techniques in historic studies of the sea and ships with the discovery and exploration of old shipwrecks. These ships have provided valuable information about how people travelled, the cargoes they carried and the methods of construction. This is just one aspect of research aided by the development of scuba diving. In the next chapter we examine ways of ensuring that the advantages of scuba exploration are not destroyed by exploitation of the environment.

The Valiant, *a tug, was being towed out to sea to be scuttled when she sank unexpectedly in 27 m of water. She has already attracted a variety of marine life which are a feature of the diving on this popular spot.*

Lady Fairfax Island has a prolific number of gannets nesting in depressions in the coral rubble. They are protected in this sanctuary and you can approach them within metres.

CONSERVING THE LAST FRONTIER

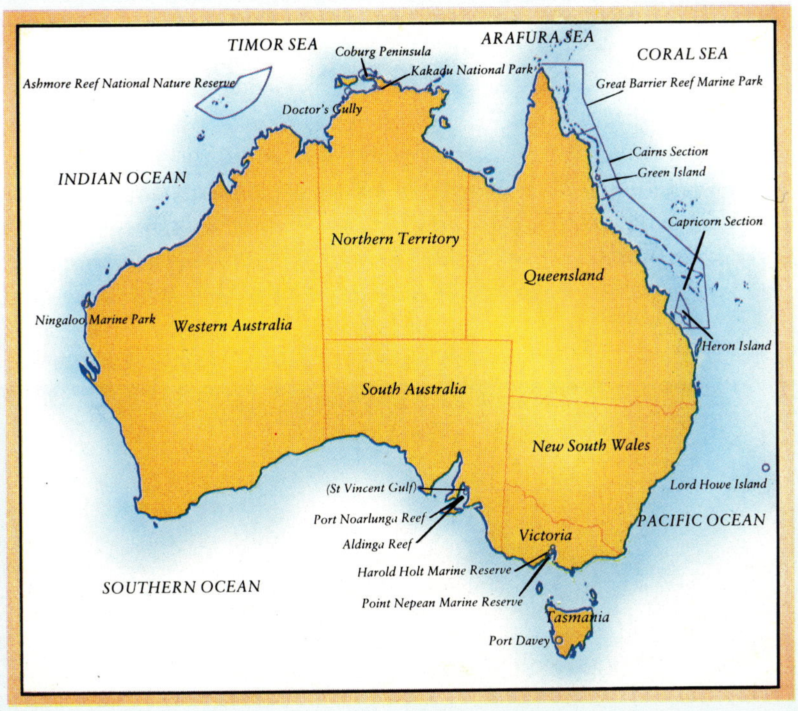

For millions of year the oceans have covered a major portion of the earth acting as a reservoir for life on land. Ninety per cent of all the earth's animals have come from the seas and today human beings utilise twenty per cent of the sea's potential supply of protein-rich food. Yet despite this, we are only just starting to appreciate life in the sea as being of value to the future. Until now we have viewed the seas as a source for easily obtainable and economic products.

Australia's increasing demand for seafood is placing an enormous strain on our fisheries, to the extent that urgent management procedures need to be taken. Also, concern over the dumping of sewerage and waste materials into the sea has led government authorities to upgrade sewerage treatment plants and ban industrial discharge into rivers. However, despite these advances, agricultural run-off is still a significant problem. (Some marine biologists already blame the abnormally high levels of nitrates and phosphates washed off Queensland farms during fierce rainstorms for the current plague of crown-of-thorns starfish on the Great Barrier Reef.)

Evidence that Australians are beginning to see their own ocean's future in clearer terms has come with the enactment of the Great Barrier Reef Marine Park Act in 1975. Marine parks ensure that whole areas — including islands, reefs, intertidal zones and open sea — are managed as a whole. Although they can be equated with national parks on the land, marine parks differ considerably in that they allow use of a region to continue with the very minimum of restrictions. It is to be hoped that Australians will quickly recognise the importance of marine conservation and management of fisheries resources.

"When I think of marine reserves I recall experiences such as these: the sight of a school of maori wrasse working their way along the reef slope in the late afternoon; as I hover above, tied to the surface by a snorkel, fish as big as cows, adults and juniors in the school, are bumping the coral with their 'foreheads'. I feel the vibration in the water, like a waterbed moving in tune to another sleeper. As I watch they pick up a piece of coral in their enormous teeth and grind and chew, spitting out little bits as they go, leaving their message of fine coral sand behind them.

I see a giant grouper in an underwater cavern. A coral cave formed at the time of lower sea level. The cave is eroded. There is a pile of rubble at the entrance, now covered with new growth. As I enter, the dimness inside the cavern gradually resolves itself into a form, a very large form, about three or four times as big as me. He is watching me, and has been all along. I move up to him, he lets me approach almost within touching distance, then with a flick of the tail he moves back further into the cave. On the roof, my bubbles are now collecting in little pools of silver. I am aware of the grouper lurking in the back of the cave but am suddenly sensitive to all the minute coral and sea fans growing from the roof. Such are the personal experiences that compel me to seek their preservation so that my children too can see and feel the magic that I have experienced."

The branching coral is growing out and up towards the light making a castle like formation as it sweeps upwards.

Australia's long coastline has a variety and beauty unequalled in the world. For over 36 800 kilometres the coast abuts various oceans — the Indian Ocean in the west, the Southern Ocean, the Pacific Ocean in the east, and in the north the Arafura, Coral and Timor seas. The enormous geographic range, extending from the tropics to the cold waters of southern Tasmania, results in a great variety of marine life. The economic potential of the continental shelf has been recognised in the declaration of the Exclusive Economic Zone (EEZ), a zone of 200 nautical miles from the coast. The resources of this area are the economic right of the adjacent country but that country also has a responsibility to ensure that exploitation does not exceed the sustainable yield of fish or other living resources. The responsibility of Australia's claim to a 200-mile EEZ, means we must be able to manage the resources of the zone. To substantiate what is the sustainable yield of a fish, or other economic resource of the area, requires marine research, and conditions such as licences, limitation of species that may be caught, regulation of seasons and areas so as to protect breeding stock and limitations in the type of fishing gear. Areas must be set aside to conserve sections of habitat in pristine condition. Thus the rationale for marine reserves is tied to a general awakening to the seas and oceans of the world, their potential and the realisation that the seas are the last frontier on earth.

The key to conserving the marine environment is the preservation of representative habitats in prime condition. In marine areas, such habitats include the underwater terrain as well as the water above it and the plant and animal life associated with both the bottom (benthic) terrain and the mid-water (pelagic) fish, and plankton which are transient occupants.

In the declaration of marine reserves, the size required may be complicated by what use the reserve will serve. Is it to be primarily for conservation, historic, scientific, educational, recreational or wilderness purposes? When the reserve is large enough, it is possible to zone areas for a variety of purposes and activities.

Conservational reserves are areas least affected by people and representative of different aquatic ecosystems. Nursery and breeding grounds for important commercial and recreational fisheries, such highly productive areas as mangroves and seagrass, are of prime importance. When they are not already in declared reserves, in New South Wales at least, development applications affecting mangrove or seagrass areas are viewed by fisheries authorities.

Above: Divers like to visit remote Barrier Reef islands such as Lady Musgrave Island in the Capricorn Group.

Below: Beachcombers enjoying a safari expedition on a remote Barrier Reef island.

How does the present system of marine reserves in Australia measure up to conservation criteria?

The first marine reserve in Australia was the reef surrounding Green Island, a small coral cay twenty kilometres from Cairns, north Queensland, which has been protected under fisheries

A pelican in the quiet calm waters of a coastal lagoon creates a feeling of peace and tranquility.

legislation since 1938. Heron Island and adjacent Wistari Reef, eighty kilometres north east of Gladstone were also protected under fisheries legislation. The declaration of the Capricornia section of the Great Barrier Reef Marine Park in 1979 and the Cairns section in 1981, under the *Great Barrier Reef Marine Park Act* has redressed the inadequacy of protection afforded our greatest marine attraction. The significance of the region has been realised with the inclusion of the Great Barrier Reef in the World Heritage Listing. The marine life of the reef comprises the diverse forms of coral: the branching corals, brain corals, plate corals, mushroom corals, and encrusting corals. There are numerous shells such as the giant clam and over 4000 other species. The fish are almost as diverse with 2000 species including the pelagic species such as marlin and mackerel, the reef dwellers like coral trout, red emperor, snapper and cod, while brightly coloured territorial fishes like butterfly fish and damsel fish defend a few square metres of reef. There are six species of turtle, the nesting sites of the green and loggerhead turtles are of world significance, while whales, dolphins and dugong also live in reef waters.

The multiple use zoning plan, restricting various activities in certain areas, enables the protection of a buffer zone, allowing traditional fishing and commercial fishing in certain areas and recreational activities in other areas. The sections proclaimed in 1983, are far northern, central, southern, inshore southern, Cormorant Pass and Townsville. In addition there are many fish habitat reserves to protect inshore and estuarine habitats in Queensland.

The area covered by the Park is certainly representative of the Great Barrier Reef, in fact with ninety-eight per cent of the Great Barrier Reef region now in the Park, the conservation of the reef is assured. Recreation and education

An egret working its way across the flats of a saltmarsh lagoon area where food is abundant.

Visitor viewing the reef through a glass-bottomed boat. A more sophisticated way to see the underwater scene now is to take a reef trip on a Sub See.

have quite different requirements. Accessibility, interesting terrain, diversity of plant and animal life, and clear water are attributes that sport divers seek in a marine park. Facilities to enable the non-diving public to appreciate the marine environment, such as glass-bottomed boats, onshore aquaria and museums, underwater observation chambers and submersible vehicles, enable them to gain an appreciation of the marine environment. For sport divers, marked scuba trails, notices indicating interesting fauna or geology and support facilities such as air compressors and even decompression chambers are highly desirable facilities.

The underwater observatory at Green Island and others operating on the reef have done much to enable visitors to appreciate the beauty of the reef, as have the glass-bottomed boats and Sub Sees operating out of various resorts. These are all private enterprise ventures, successful in promoting the reef in ways that are inaccessible to public administrators. Encouragement of private enterprise offering facilities for the public that enhance appreciation of the reef becomes a desirable feature of marine park planning.

No marine parks in Australia have been specifically designed to enhance the recreational or educational potential as has been done in Japan or the United States. The Kushimoto Marine Park on the Kii Peninsula, at the southernmost point of the Japanese island of Honshu, is headquarters of the Marine Parks Centre of Japan and its facilities include a large onshore aquarium, underwater observation towers that hold forty people at a time, and glass-bottomed boats that carry as many people on regular tours of the park's coral reefs. We do not have the system of marked scuba diving trails such as in the John Pennekamp Coral Reef State Park in the United States of America, yet with the finest coral reef system in the world, we have enor-

The swimmer is enjoying the tidal pools on the edge of the coral reef.

A pair of raccoon butterfly fish amongst coral growth at the Abrolhos Islands, 70 km offshore from Geraldton, WA.

A feather duster worm with gills extended in the water will quickly retract if there is any movement in the water nearby.

mous potential to increase the tourist and educational facilities provided.

In Western Australia, several areas have been proposed as marine reserves; fortunately, the West Australian National Parks and Wildlife Service now has a legislative provision to declare them. The marine reserve of Ashmore and Cartier Islands, in the Indian Ocean, is representative of one biogeographic area. It has been gazetted under Australian National Parks and Wildlife legislation as Ashmore Reef National Nature Reserve. The reefs and islands comprise an external territory of Australia that extends our marine boundaries. Indonesia is disputing our rights in the area, claiming traditional fishing rights — Australia claims the Timor Trough forms a natural boundary; Indonesia claims a boundary equidistant from the two shores. An *Agreement of Understanding* between Australia and Indonesia gives Indonesian fishermen access to two coral cays but there have been transgressions of the agreement. Indonesians have been camping on the Islands and killing seabirds and turtles. The navy is patrolling the area and health and quarantine officials have issued warnings. A Marine Parks Act has been passed to supplement the present legislation.

The Ningaloo Reef is proposed as a marine reserve including 200 kilometres of coastline off North West Cape. It represents a coastal coral reef system, averaging 2.5 kilometres from the coast, stretching for 260 kilometres. The proposed Ningaloo Marine Park lies just north of Point Murat and extends around North West Cape to Amherst Point, Western Australia. The marine park will extend offshore for approximately ten nautical miles from the low tide mark, encompassing an area of 5000 square kilometres. It lies just offshore from a very arid part of Australia. The reefs are typical of Indo-West Pacific fauna but there are also temperate species at the northern limit of their distribution in Western Australia. Turtles are common, with several rookeries along the coast, and whales visit the area. The area is popular with tourists, with commercial development centred at Exmouth and Coral Bay. The recommendations for the establishment of the Park propose three major zones: a General Use Zone where commercial charter fishing is allowed, under permit — fish caught are not to exceed the limits of reasonable personal consumption with a maximum weight limit; professional fishing is allowed only outside the recreation zone and by permit; a Sanctuary Zone; and a Recreation Zone with areas subject to seasonal closure to protect breeding stock, are proposed. The management agency, probably Fisheries and Wildlife and National Parks, propose to provide recreational facilities — snorkel trails, scuba trails and nature and reef walks in liaison with the Department of Tourism. Tenure of the marine areas of the proposed marine park is vested in

Typical gorgonia sea fan found on the reef. They are filter feeders preferring the walls and cave entrances where there is water flow and they are shaded from direct sunlight. They grow out from the surface of the reef.

The Lord Howe Island woodhen is a flightless bird which was killed for food in the early days of settlement on Lord Howe Island. Cats, dogs and rats have preyed upon them in more recent years until their numbers were reduced to below 20. On Mt Gower are small canyons in a remote area where the woodhens survived in small numbers. On this trip National Parks and Wildlife Service officers abseiled into the area to assess the numbers. Dick Smith's helicopter was used to transport woodhens to a compound at lower levels where they have been successfully bred in captivity rescuing them from the verge of extinction.

several State and Commonwealth bodies. Part of it is a military training area. Rowley Shoals, Shark Bay and Broke Inlet are also awaiting suitable West Australian legislation to be passed. A problem has been the lack of Commonwealth legislation for marine reserves that would ensure a national management philosophy. Now that a Marine Park Bill has been implemented in Western Australia, this reserve has the potential for very imaginative management as a recreational resource.

Rowley Shoals is a much more remote area about 300 kilometres north west of Broome. Ashmore Reef is a northern extension of the shoals. Rowley Shoals, which lie on the edge of the north western Australian shelf, are perfect examples of shelf reefs and atolls. Each has a wide reef rim enclosing a lagoon or series of lagoons broken by narrow passages, with a tidal raceway on the north eastern side. They have an unusual tidal range of about five metres, that races through the reef passages. At low tide the water drops about a metre lower outside the lagoon. The water in the lagoon is above sea level. On their western edge the bottom drops rapidly into a deep ocean basin (4000m). Charter boat services have been operating out of Broome in winter since 1977. Diving the sheer drop-offs, sheltered coral lagoons and tidal raceways is some of the most spectacular diving in Australia. This area is also awaiting suitable West Australian legislation.

Lord Howe Island, the southernmost coral reef in the world and an area of great zoogeographic interest with particular marine importance, is proposed as a marine reserve though areas in the Admiralty Islands have not been included. The Solitary Islands, off Coffs Harbour, New South Wales, with its interesting reef-building corals and mixture of tropical/temperate fauna is also proposed. New South Wales has a fairly good record with the establishment of marine reserves beginning with the proclamation of a marine extension of Bouddi State (now National) Park in 1971. An amendment to the *Fisheries and Oyster Farms Act* in 1979 enabled the establishment of aquatic reserves. Aquatic reserves close to Sydney are at Long Reef, Bass Point, Shiprock, and in the northern part of Sydney Harbour.

The Long Reef Aquatic Reserve was gazetted when it came under pressure from people collecting shellfish and sea urchins for food. It has been used as a field studies area for local schools and universities for over fifty years. Limited collecting is allowed under permit from New South Wales Fisheries Division.

Shiprock is well known to divers around Sydney. It is a deep submarine cliff where strong currents supply oxygen and nutrients creating a profusion of marine life. Sponges, ascidians, bryozoans, hydrozoans cover every available space with species which are rare elsewhere. It was declared an aquatic reserve in 1979. The taking of all marine life for a distance of sixty metres from the shore is prohibited. Since Shiprock is in the Sydney suburb of Lilli Pilli and there are suburban residences on the foreshore, access is difficult and a continuing problem with no local government interest in providing steps on the only shore access point. North Head to Little Manly Point and between Grotto Point and Forty Baskets Beach in Sydney Harbour is a marine reserve with some restrictions on fishing and removal of marine life.

Near Shellharbour on the New South Wales south coast is the Bushranger Bay (Bass Point) reserve, an ideal beginner diver area protected

Coconut palms conjure visions of tropical islands but on Australia's tropical islands coconut palms are rare. These are in fact on subtropical Lord Howe Island.

The spectacular starfish has hard multicoloured lumps on it which give it a plastic feel when handled.

Close up of the starfish's beautiful patterned skin.

Neds Beach on Lord Howe Island is a popular spot at mealtime. Many pelagic fish as well as reef dwellers come right up to the swimmers and take food from their hands.

against the removal of marine life. There is a cave at Bushranger Bay hidden in the walls of a vertical underwater cliff, and the drop offs on the island are interesting. There are many wrecks in the area, Bass Point was a local shipping centre before the railway took over.

On the north coast, near the Queensland border, is Julian Rocks, located off Byron Bay. It is an area supporting an interesting mixture of tropical and temperate water marine life which is protected for a distance of 500 metres from the trig station on the Rocks.

The first marine reserve to be declared in Victoria was the Harold Holt Marine Reserve. The reserve is scattered in discontinuous areas and not well known locally. When I approached the Harbours and Marine authorities in the Queenscliff port area and asked if they could indicate where the marine reserve was, they did not know. It isn't signposted, though there is a marine research station nearby. Sections are restricted to military personnel by guarded bridge.

The Point Nepean Marine Reserve on the other side of the Port Phillip Heads has entry restricted to occasional supervised trips to Cheviot Beach where former Prime Minister Harold Holt disappeared. Swan Bay, on the Queenscliff side, has some of the last remaining seagrass habitat in Port Phillip Heads providing an important fish and waterbird nursery area, but commercial fishermen can net outside the area and angling with rod and line is permitted. Amateur fishermen are not allowed to take abalone or crayfish from any part of the Harold Holt Marine Reserve but licensed fishermen are permitted to do so, at Point Lonsdale and Point Nepean. This is the area where restricted entry to the military reserves provides the greatest conservation protection. Sea birds such as storm petrels, caspian terns, migratory waders and silver gulls are found on the sea grass and the salt marshes of Mud Island that is a nesting area from September to December. Australian gannets nest at the Wedge Light while Evans Point and Duck Island near Swan Bay are wading bird habitats. The Mud Island Marine Reserve is open to all forms of traditional fishing including spear fishing.

Pope's Eye, a semicircular reef enclosing a lagoon, is a man-made fort in the centre of Port Phillip channels. It was constructed on Pope's Eye Shoal as a fortress but was never completed. The semicircular ring of bluestone forms an anchorage for spectacular marine life, fed by strong currents supplying plankton and nutrients. The central lagoon can be dived regardless of the tide, unusual in Port Phillip where divers must wait for slack tide to enter the water, at most of the popular locations. Pope's Eye is one of the most popular diving spots in Port Phillip and is protected from all forms of fishing.

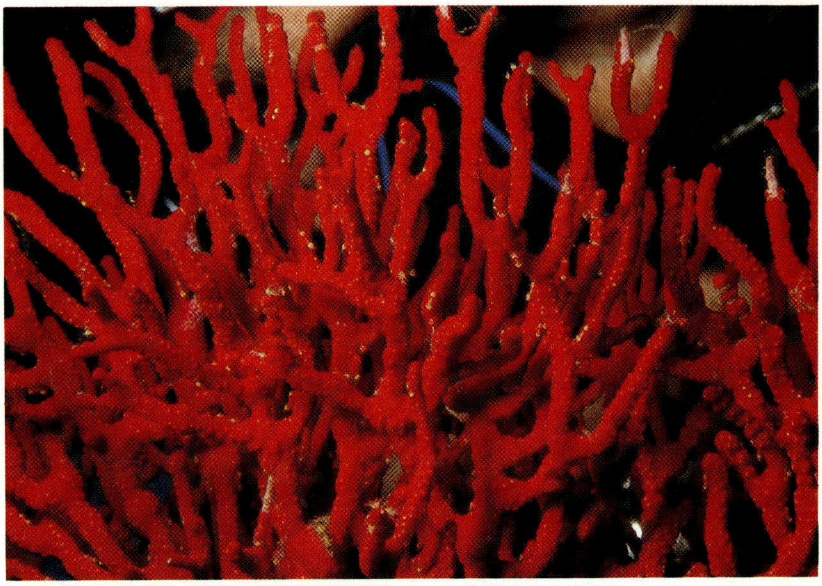

Above: Small gorgonian coral is a feature of caves in southern Australian marine reserves.

Below: Pope's Eye in the middle of Port Phillip is a man made reef which divers favour as it is diveable in all tides and has interesting marine life. This fish eye view shows the dive boat anchored in the calm central lagoon.

The basket star is a many armed sea star which entwines itself on the gorgonia extending its branched arms to filter feed. In the day it is curled up tightly in the gorgonia but in the late afternoon spreads its arms to feed.

Management of the Harold Holt Marine Reserve suffers from its scattered locations and the conflicting traditional land uses that have requirements not compatible with conservation and recreation. The inclusion of the headlands as part of the marine reserve and a total management of land and marine areas would be a more rational approach.

South Australia has several Aquatic Reserves under its *Fisheries Act*, declared since 1971. Close to Adelaide is the Port Noarlunga Reef in St Vincent Gulf, a sandstone reef about 1.5 kilometres long connected to the shore by jetty. Port Noarlunga was once a thriving port where barges brought cargo down the Onkaparinga River. An anchor from the sailing ship, *David Witton*, that sank on the reef in 1838, is displayed in the lifesaving complex beside the jetty. Fish populations were decimated by spearfishing in the early 1960s, but since its protection, fish are plentiful and unafraid of divers. The slower growing gorgonians that once adorned the caves and ledges, and the black cowries that once appeared at certain times of the year to lay their eggs are now very rare. It is an example of an educational, recreational reserve and a popular South Australian diving spot. Ten kilometres south is Aldinga Reef, an intertidal limestone platform primarily a recreational/educational reserve with precipitous underwater cliffs and caves. Goose Island and some nearby islets in Spencer Gulf are primarily an educational reserve for an Adelaide school. There are several other conservational reserves including Barker Inlet on the Torrens River in South Australia, one of many of the State's mangrove areas that have been protected by fisheries legislation. On Kangaroo Island is the American River Aquatic Reserve, an estuarine nursery area and Seal Bay Aquatic Reserve, primarily a conservational reserve to protect the Australian hair seal.

There are several marine reserves in the Northern Territory. They are the Coburg Peninsula, Kakadu National Park and Doctor's Gully.

In Tasmania, a proposal to have Ninepin Point Reef in D'Entrecasteaux Channel near Hobart gazetted, resulted in exploitation of the reef by fishermen with resulting depletion of fish stocks.

Port Davey is the southernmost part of Australia which is a declared marine reserve. It is a drowned river valley in south west Tasmania. This area is one of the most remote in Australia, the population consisting of two families of tin miners and a few bushwalkers and fishermen.

There are difficulties in the present system of marine reserves between State and Commonwealth jurisdiction. An overall management strategy is essential if we are to learn from our earlier mistakes. Marine areas, with their dependence on outside inputs to the ecosystem, need large scale management. It is impossible to conserve our fish stocks if we are destroying the breeding areas. When mangroves and salt marsh are filled, tidal flats dredged, watercourses silted by poor land management, then we are destroying the source of marine productivity.

In Queensland, the Great Barrier Reef Marine Park operates in joint management between the Great Barrier Reef Marine Park Authority and Queensland National Parks and Wildlife Service, who provide rangers and field staff. In the Capricornia section of the Park there are about ten field staff working out of Rockhampton. To manage and implement policies over such a large area, obviously requires more staff than is available at present.

Mangroves are evergreen trees that grow on the coastal fringe between land and sea and help to stabilise the delicate balance between them. The roots of mangroves trap sediments brought down by rivers while the leaves provide the basic food for many animals living in the estuaries. Bacteria and detritus feeders (worms and small crustacea) break down the leaves releasing its nutrients for other animals in the estuary. Birds rear their young on the fish and crabs living in the shallow waters. Currents and tides carry nutrients to offshore waters, where other animals, small fish and birds benefit from the mangroves' productivity.

Estuaries and mangroves are some of the most productive ecosystems in the world. Their primary productivity may be ten times that of forest and grasslands. They are very efficient users of solar energy, much more efficient than man is in agricultural production. But how can mangroves be so productive when we do not see any use for them? When their leaves fall, they are distributed by tides and currents over wide areas of the estuary. Here they are eaten by microscopic animals and bacteria which release nutrients and detritus for other small inhabitants like worms, crabs, prawns, snails that in turn are eaten by small fish, larger crabs, birds and prawns, much of which finds its way to man. In fact eighty per

Gorgonian corals and soft corals are both filter feeders but the soft coral expands in size as it takes in water while feeding. Light shafting through the water behind the diver emphasises the beauty of the underwater scene.

Sea tulips are admired by diver Steve Barnes in Jervis Bay. Their colour varies as it comes from a sponge growing over the outside coat.

Two magnificent butterfly cod hovering in a deep cave at Lady Musgrave Island. The diver entering the cave is illuminated by a second slave strobe light which fires at the same time as the close up strobe. The cave is 25 m at the end of the reef wall where it meets a gently shelving slope.

The Lord Howe Island double headed wrasse and a spangled emperor are hand fed at Comet's Hole, Lord Howe Island.

Mangroves in a quiet backwater with the mountains in the background.

cent of fisheries in Australia are based on production from estuarine and shallow marine environments.

Mangroves occur around the north Australian coast from Shark Bay in Western Australia to Darwin then round to Port Douglas and the Daintree River on the east coast. Here they attain their most complete development of zones and greatest number of species. They occur sporadically down the east coast as far south as Westernport Bay near Melbourne and in the South Australian Gulfs where they are thought to be relics of warmer conditions.

Mangroves grow along the shores of estuaries where they intermediate between sea and land, trapping the sediments washed down by creeks and rivers and using the power of the sun to make food for the many animals and plants of the sea that depend on them for food and shelter. They are adapted to the high salinity of their environment that may be over ninety per cent in the ground water around their roots. They either exclude the salt or excrete it through glands in their leaves. Salt glands enable some mangroves to secrete salt, while succulence is a method of salt avoidance used by some salt marsh plants. Some plants shed leaves, thereby getting rid of excess salt. Other salt marsh plants get rid of salt from their leaves through special salt glands. Some other plants largely exclude salt from the water they take up through their roots, thus avoiding the problem. Whatever mechanism is used, the presence of salt in varying amounts is an important environmental factor in determining the distribution of different species.

In northern Australia, where the mangrove species have reached their greatest diversity, they are zoned responding to variations in soil salinity, intertidal level, fresh water and drainage. The most tolerant of changes in salinity is the

Seagrasses form in sheltered estuarine waters and are an important habitat for breeding fish and water birds. They are highly productive areas which should be protected from dredging, rubbish dumping and indiscriminant destruction.

species that also occurs in the South Australian Gulfs. Closest to the sea are the pioneers mangroves. Next is the zone extending also along the waterways, with their tangled arching roots and dark green leaves. The tallest mangroves, often thirty metres high, form a closed forest in the next zone with species such as cannonball and cedar mangroves or looking glass mangroves also present. The fourth zone of spurred mangroves form thickets where the soils are drained. Another type indicates the presence of freshwater and occurs along the margins of freshwater swamps or the banks of tidal rivers further upstream.

Between the mangroves and the coastal forests, lies a dry land zone dominated by salt marsh plants that form the ecological association known as saltmarsh. In southern Australia, saltmarsh largely replaces mangroves occupying similar environments. In northern Australia, fluctuations in sea level have left a series of raised beaches and a barrier spits running parallel to the coast. Sometimes these lows, between the barrier spits, have freshwater ponds with water lilies, reeds and sedges. Alternatively they may be salt pans, or colonised by mangroves or saltmarsh plants. In the absence of barrier spits, estuaries run to the sea fringed by extensive flats of muddy sand and mangroves. Mangroves will grow on sand but once they are established, mud accumulates, often overlying the original sand. Mangroves are able to render calcareous rock soluble and can occur on offshore coral islands, particularly on their southern ends. The mud associated with mangroves is mainly organic from their own making and calcareous, but also comes from silts brought down by heavy flooding. Offshore from the mangroves where the water is clear, seagrass beds extend from the mangroves seawards.

Large areas of mangrove and saltmarsh in Australia as elsewhere in the world have been reclaimed for sugar growing, industry, rubbish dumping, playing fields, shipping facilities. Many other areas are endangered by pollution of our waterways. Mangroves are not wastelands, they are important productive habitats essential for the survival of our marine environments. Economic measurements of the value of production from mangrove areas are difficult given the complex interaction of factors but one study in Moreton Bay estimated an acre of mangroves produced $300 per year in shellfish and fish with a total value of $12 million. In Florida mangroves are estimated to generate $4000 per acre of fishing tourism and recreation each year. In Australia national parks legislation stops at high water mark and fisheries legislation has often been inadequate to prevent destruction of this important marine resource. Once again, a national policy would help protect the fringe of the sea.

Seagrass beds are dominated by two species of flowering plants that grow in shallow marine environments down to seven metres in estuaries in temperate Australian waters. They are most important commercially which was discovered when 'wasting disease' hit the seagrass beds of Europe and America with the consequent failure of various fisheries in these countries. Associated with the seagrass, *Zostera*, are crustaceans such as prawns and shrimps, worms, shellfish and a large flora of microscopic algae which serve as food for mullet and other plant-eating fish. Copepod crustaceans harbour in its fronds attracting small fish that shelter from their enemies. Garfish eat the *Zostera* and larger fish eat them, while swans feed on the seagrass beds often denuding large areas. Once the bottom sediments are no longer held in place by the *Zostera*, the currents and winds may move the bottom sediments, oxidising them and making them unfit for further growth of *Zostera*, which

Above: Shiprock is a prolific marine drop off in Port Hacking estuary, Sydney. It is a declared marine reserve because of its amazing variety of marine life in a very small area close to residential and heavily populated areas.

Below: Plankton, like other algae, can multiply rapidly when conditions are right, forming ugly plankton blooms which if in enclosed waters may smell when they die.

The river has reached the sea and deposited silt out to sea, wave action has sorted the sand and ripples reflect the coastal topography and direction of the prevailing wind and currents.

prefers reducing conditions. *Posidonia*, a wider strapped seagrass, often replaces *Zostera* particularly where the bottom is sandy and reaches its greatest development in the South Australian Gulfs and the west coast of South Australia where it is so abundant that it has been commercially harvested for fibre at various times. On the east coast, it occurs in association with *Zostera*. Its replacement of *Zostera* is thought to be the result of man's interference by dredging and digging for worms to which *Posidonia* is more tolerant. The commercial rock oyster is closely associated with *Zostera* beds and the community in the seagrass provides micro-organisms that the oyster requires as food as well as nutrients such as sulphur, phosphate and nitrogen that are important to the oysters. Sewerage discharge can be a major factor in the death of seagrass because the abundance of nutrients causes epiphytic algae to accumulate at such a rate that the leaves of the seagrass *Posidonia* are weakened and drop off. This was implicated in the die-back of seagrass at Glenelg in South Australia. Removal of sewage further offshore would move, if not solve, the problem.

Eutrophication, or enrichment of the waters by the addition of excess nutrients, leads to excessive algal growth. In nature, this enrichment leads to 'red tides', explosive growths of microscopic algae or dinoflagellates which colour the sea red or green, often over vast areas. This is the result of natural factors, possibly nutrient enrichment through upwelling combined with calm sunny conditions. Colours of the bloom go from green to yellow, amber brown or red. It is during the red phase that the cells are limited in nutrients and die — releasing toxic substances that can kill large numbers of fish. Eutrophication also occurs as a result of pollution, particularly pollution of enclosed

Jellyfish float or swim in the water column, moving in diurnal rhythms, a miniature world with their associated fish or crabs in their tentacles.

Divers explore the underwater wilderness around Lord Howe Island.

bodies of water such as coastal lakes. Algae are able to store excessive amounts of phosphorus and release it as soon as they stop growing, thus making phosphorus available to the younger cells. This tends to make the algae less dependent on the soluble phosphorus in the surrounding waters during bloom growth periods. The same phosphorus may be used two or three times in producing the algal bloom and is fuelling the cycle. Control of the biodegradable organic matter combined with removal of nitrogen is a more satisfactory means of limiting the nuisance algae. Another source of environmental stress is the cycling of cooling waters through power stations. For every kilowatt of electricity generated, there are two kilowatts of waste heat to be dissipated. The temperature tolerances of aquatic organisms can be exceeded by thermal discharges and have been associated with increased incidence of fish diseases.

Coastal zones are extensive, valuable resources but coastal erosion is a worldwide problem. The offshore reserves of sand that fed the increasing width of beaches between 4000 and 6000 years ago is now depleted. There are local variations where sediment is being generated from extensive cliff areas or being brought down by rivers. The evidence is, that most of this is being deposited in coastal lakes and lagoons and not reaching the depleted offshore areas. A stable foredune or first dune acts as a natural protection to the coast against wave erosion. It is a buffer circulating sand between dune, beach, surf zone and sea bed. As long as there is sufficient volume of sand, the foredune will be able to be eroded by storm waves, to be replenished in calmer conditions. It acts as a barrier to protect the landward zone from winds, salt

Above: Funds for beach improvement are sometimes misused by local councils. At Palm Beach isthmus, where parabolic sand dunes had developed, Warringah Shire Council flattened the dunes to plant spinifex in the mistaken belief that the dunes were in danger. The parabolic dunes were a mature development of a hind dune system.

Middle: Saltmarsh and hind dune plants use their succulent leaves and salt secreting ability to adapt to the dry saline conditions.

Below: These saltmarsh and hind dune plants are pictured along the coastline near Sydney.

Clown fish are common in the lagoon of Lord Howe Island. This one aggressively defends his territory around the sea anemone.

Above: Dredges such as this, make their own pond on which they work the sands of the hind dunes for minerals. Mineral sand mining leases have been granted for large areas of our east and west coasts.

Below: Pollution caused by nutrient enrichment with subsequent algal bloom growth, is common in urban coastal lakes.

spray and sea water flooding. A break in the foredune leads to funnelling of wind and blow outs, where sand is moving inland unprotected by vegetation cover. This is a natural occurrence in undeveloped areas and the dune finds its optimum stability but in developed coastal areas where buildings have been constructed too close to the sea this can have disastrous results. Houses, swimming pools and jetties may topple into the sea in storm wave conditions. Investment by local government in protecting seashore properties, because of unwise development decisions in the past, is enormous. Ultimately we are paying for past mistakes, yet such development continues all along Australia's coastline. Councils continue to approve development applications in the nearshore coastal zone, and continue to spend money on beach improvement, building groynes or sea walls, even flattening sand dunes to plant spinifex. Flattening the nearshore dunes at Sydney's Palm Beach was a classic case of misuse of government funding for beach improvement. The parabolic dunes were well developed by wind, yet the local council 'improved' them by creating a flat expanse of neat rows of spinifex grass.

Sand mining is common in coastal areas along the east coast of Australia where minerals such as rutile, zircon, ilmenite and monazite occur along old strandlines and dunes, visible as dark coloured sand layers in the lighter sands of our beaches. They occur there as a result of earlier weathering and transportation, then are concentrated by wave action during heavy storms, or by wind that blows the minerals landward to form dunes in the direction of the prevailing wind. The minerals are mined either by an excavator or front-end loader which is dry mining, or by a hydraulic section-cutter dredge that cuts below the water table and pumps the mineral sand to a concentrator. The concentrator separates the heavy mineral by its specific gravity, which is greater than the sand, and final separation is by electrostatic and electromagnetic means.

There are three major areas of sand mining activities in Australia: the east coast deposits from Sydney north to Fraser Island; the south western deposits between Bunbury and Busselton; and the Eneabba area north of Perth. Mining licences are granted by the State governments which presents difficulties when conservation issues are being considered. In 1976, beach sand mining on Fraser Island, the largest sand island in the world, was destroying its unique rainforest. D. M. Minerals and Queensland Titanium Mines Pty Ltd had mining leases on the island and were selectively mining and replanting. However since it takes 300 years for a mature rainforest to reach its climax condition or maturity, their efforts were somewhat inadequate.

Sand blowouts occur where sand dunes have been stripped of vegetation or on dunes which are moving.

A political furore between State and Commonwealth governments ensued in which the federal government resolved the issue by cancelling the export licence. A compensation claim by

Chris Deacon is observing two crinoids or feather stars. She is wearing a pink lycra suit for protection from cuts in tropical waters.

DACOR
AQUASTROBE 100

Island nations have signed fishing agreements with other nations for a percentage of the fish caught in their waters. In the Solomon Islands this Taiwanese vessel is operating with a Solomon Islander crew.

the mining companies followed. Where there is conflict of options for land use, a rational approach is to zone areas for different activities. Zoning over a period of time can allow non-compatible uses. For example, mineral sand mining may be followed by replanting and residential development or recreation.

The important consideration is planning for the future so that unnecessary destruction of the environment is prevented and the limited coastal resources are used to benefit the maximum number of people over the long term. Conflict can only be resolved by comprehensive management plans and clearly defined State and federal government policy. Cost-benefit analysis provides a framework for evaluating proposed coastal land use but has difficulties in assessing environmental values. Unique or unusual natural features may be lost by development. For most people, the preservation of the natural environment adds to their enjoyment of life, but the problem of estimating the value of land kept in its natural state makes it difficult to compare the relative merits of land use proposals. Bias towards commercial ventures can be expected since it is difficult to justify keeping land in its natural state, when the alternative benefits of commercial development can be more accurately quantified. It is becoming increasingly necessary to place commercial values on coastal and estuarine areas which emphasise their recreational, tourism, educational and fisheries values in order to retain coastal areas in their natural state, yet very little work in cost-benefit analysis is undertaken by scientists or management bodies.

The declaration of the 200-nautical mile Exclusive Economic Zone gives Australia jurisdiction over an enormous area. Sovereignty under the law of the sea gives Australia the resource rights to the living and mineral resources of the seabed and continental shelf with the responsibility to conserve and manage these resources. The *Fisheries Act* 1978 gives Australian fishermen the rights to fish in the 200-mile Exclusive Economic Zone and foreign fishing vessels may fish only by licence. Licences are granted on the basis of the sustainable fishing yield, yet in many cases we have no idea what fishing pressure can be sustained, for we have no information on the population dynamics of many important commercial fisheries. While the Australian fisheries catch is small by world standards, the annual value is relatively high because of the emphasis on the valuable prawn and crayfishing industries. Less than ten per cent of fish species are commercially exploited. More research into fisheries yields and diversification of the fisheries market would enable us to develop resources we are not currently exploiting.

Although Australia has first right of access to the Exclusive Economic Zone, it also has an obligation to permit other countries to utilise this resource if Australia is not capable of utilising the maximum sustainable yield. Joint feasibility studies with other nations aim to develop resources that Australia is currently not exploiting. Fish consumption in Australia is six kilograms per person per annum and crustaceans and molluscs account for 2.3 kilograms per person per annum. Tuna is the main fish product followed by mullet, shark, gemfish, redfish, Australian salmon, snapper and whiting.

Administration of fisheries has a number of objectives, the most important being to conserve living resources and to sustain the maximum yield consistent with economy in exploitation. Fisheries resources are common property. Any restrictions on the numbers of boats operating in a fishery are designed to maintain adequate fish stocks and provide a viable living for those fishermen financially dependent on a particular fishery. Other management measures that have

Staghorn coral forms massive branching structure in calmer tropic waters.

Above: A spoonbill is reflected in the quiet waters of a coastal lagoon, an important breeding habitat for this and many other birds.

been introduced include closed seasons to protect the breeding stocks, minimum size limits, closed areas and regulations of the types of fishing gear.

Such measures provide an example of those we need to take in order to conserve our marine environment. I am reminded of my study of a local coastal lagoon at Curl Curl. When I read the historical reports of the early settlers sliding down huge sandhills into a lake that abounded with waterfowl, I thought they were speaking of a different lagoon. The degraded wasteland of football fields and basketball courts, the smelly rubbished salt marsh and the polluted drains that emptied into the upper reaches bore no resemblance to the reports I was reading. Yet this degradation, created in part by council rubbish dumping with public support and cooperation, has happened within living memory. We must conserve not only what remains in an unspoilt state, but try to redress past mistakes in water and land misuse by a philosophy of concern for the environment. This comes down to a personal, practical level of not washing the car in the street where nutrients travel down the stormwater drains providing conditions for algae to increase in blooms in our rivers and lakes causing rotten egg smells and unsightly scums. It means throwing back that undersized fish, refusing undersized crayfish in restaurants, it means not collecting live shells, or coral, or aquarium fish — it means caring for what we love.

In this book we have attempted to show you the beauty and wonder of what lies in Australia — down under. I hope that the barren wasteland I have just described will be a very uncommon sight, and that after reading the book you will be moved, as we are, to protect the sea for our children. We need careful management of our marine areas to attain a balance between development and conservation. The value of our marine resources lies not only in commercial development, but in many tourist and educational opportunities. Understanding and an appreciation of the sea leads to a desire to protect it from destruction. Scientists and divers and members of the public must work together to increase knowledge and teach others to appreciate the sea and its life.

Marine research instructor, Neville Coleman, points out features of the reef to students Scott and Joanne who were taught to dive as part of a rehabilitation program.

The quiet waters of Lady Musgrave Lagoon reveal the beauty of a coral reef.

APPENDICES
DIVE SHOPS

New South Wales

ACTION DIVERS: 37 Head Street, **Forster** NSW 2428

ACTION DIVERS: Jim Holiday, 410 Maitland Road, Mayfield, **Newcastle** NSW 2300

ACTION DIVERS: 70 High Street, **Swansea** NSW 2281

ALAN JARETT'S DIVER'S WORLD: Alan Jarrett, 23 Cherry Street, **Ballina** NSW 2478

AQUASPORTS (YAGOONA) PTY LTD: Tom Byron, 430 Hume Highway, **Yagoona** NSW 2199

ATLANTIS DIVERS — BARRENJOEY B/S: Mike Buesnel, Governor Philip Park, **Palm Beach** NSW 2108

BAY AQUATICS AND ALLSPORTS: Keith Hughes, 1–2 Cinema Mall, **Nelson Bay** NSW 2315

BLUE MARLIN DIVERS: 126 Princes Highway, **Ulladulla** NSW 2539

BYRON BAY DIVE CENTRE: Bill Sylvester, 9 Lawson Street, **Byron Bay** NSW 2481

CAMPBELLTOWN DIVE: Robyn Watson, Shop 1 Cinema Complex, Dumaresq Street, **Campbelltown** NSW 2560

COFFS HARBOUR DIVE CENTRE: Phil King, PO Box 395, **Woolgoola** NSW 2456

CRONULLA DIVE SHOP: Dave Ireland, Shop 4 Kingsway Arcade, Kingsway, **Cronulla** NSW 2230

DARRYL'S FISHING AND DIVE SHOP: Darryl Stuart, PO Box 265, **Narooma** NSW 2546

DEEP 6 DIVING: Marinka Rosenfeld, 1091 Victoria Road, **West Ryde** NSW 2114

DESCEND DIVER EDUCATION: Des Walters, 810 Delaney Street, **Albury** NSW 2640

DIVE ABOUT: The Manager, Unit 6 Moruya Plaza, 62 Vulcan Street, **Moruya** NSW 2537

DIVE TREK: 153 Bayswater Road, **Rushcutters Bay** NSW 2011

DIVERQUEST: Steve Moody, Lot 348 Mullaway Drive, **Mullaway** NSW 2450

DIVE 2000: Kevin Deacon, 2 Military Road, **Neutral Bay** NSW 2089

FATHOM DIVING: Ted Lauger, 178 Sydney Road, **Fairlight** NSW 2094

FORSTER DIVE SCHOOL: Ron Hunter, Fisherman's Wharf, **Forster** NSW 2428

FORSTER DIVE CENTRE: 13 Little Street, **Forster** NSW 2428

FROG DIVE MERRYLANDS: Bob Taylor, 7A Miller Street, **Merrylands** NSW 2160

FROG DIVE SERVICES: Josh Martin, 1/20 Castelraigh Street, **Penrith** NSW 2750

FROG DIVE SERVICES PTY LTD: Peter Eason, 35A Albany Street, **Crows Nest** NSW 2065

FUN DIVE: Bob Cason, PO Box, 136, **Petersham** NSW 2049

GOSFORD DIVER SERVICES: Jim Edwards, 310 Trafalgar Avenue, **Umina** NSW 2257

HYDRA DIVING: Barry Copelin, 16 Florida Street, **Port Macquarie** NSW 2444

ILLAWARRA AQUA CENTRE: Les Griffin, 312 Windang Road, **Windang** NSW 2503

J B SEA SPORTS: Adrian & Pat Cookson, 49 Owen Street, **Huskisson** NSW 2540

MID COAST DIVERS: Mike Emery, The Marina, Park Street, **Port Macquarie** NSW 2444

MID COAST DIVERS — ST IVES: Steve Gribble, 4/194 Mona Vale Road, **St Ives** NSW 2075

MID COAST DIVERS — PORT STEPHENS: Steve Mackaway, Public Wharf, Ternby Road, **Nelson Bay** NSW 2315

MID COAST DIVERS — SOUTH WEST ROCKS: Bill Wall, Costa Rica Motel, **South West Rocks** NSW 2413

MIRANDA DIVE: Phil Challinor, 7 Beach Park Avenue, **Cronulla** NSW 2230

MONTAGUE DIVE SERVICES: The Manager, PO Box 28, **Bermagui South** NSW 2547

NEWCASTLE DIVING ACADEMY: Frank West, 430 Pacific Highway, Belmont, **Newcastle** NSW 2280

OCEAN DIVERS (NSW): Marco Kottakis, 87 Beaumont Street, **Hamilton** NSW 2303

PHIL STRAW'S DIVE CENTRE: Phil Straw, 38 Kingsway, **Cronulla** NSW 2230

PIONEER DIVING: 355 Clovelly Road, **Clovelly** NSW 2031

POSEIDON DIVE: Sue Kafe, Clydeside Pier, **Batemans Bay** NSW 2536

PRO-DIVE SERVICES: The Manager, 59 Donald Street, **Nelson Bay** NSW 2315

PRO-DIVE SERVICES: The Manager, 22 Victoria Road, **Drummoyne** NSW 2047

PRO-DIVING SERVICES: Rick Poole, 27 Alfreda Street, **Coogee** NSW 2034

PRO-DIVING SERVICES: Geoff Shelton, 76 Owen Street, **Huskisson** NSW 2540

REEF DIVE: Anthony Shelley, 322 Pacific Highway, **Hornsby** NSW 2077

ROCKS DIVE 'N' SURF: Bob Sands, 15 Glandstone Street, **Kempsey** NSW 2440

ST GEORGE UNDERWATER CENTRE: Paul Missingham, 456 King George's Road, **Beverley Hills** NSW 2209

SAPPHIRE COAST DIVERS: Maurie Vierow, PO Box 313, **Merimbula** NSW 2548

SCUBA WAREHOUSE: Judith McDonald, PO Box 1280, **Parramatta** NSW 2150

SCUBAROO DIVE SHOP: Geoff Sumner, 36 Melbourne Street, **East Maitland** NSW 2323

SHIPROCK DIVE: Wayne Christiansen, 709 Port Hacking Road, **Donans Bay** NSW 2228

SNORKEL INN: Sid Millott, Shop 4 Moorefield Concourse, **Kogarah** NSW 2217

SOLITARY ISLANDS DIVE: Andrew & Carol Wright, 396 High Street, **Coffs Harbour** NSW 2450

SOUTH WEST ROCKS MARINE: 100– 104 Gregory Street, **South West Rocks** NSW 2431

SOUTHSIDE DIVER TRAINING: Sue Sainsbury, 29 Warrah Road, **Yowie Bay** NSW 2228

SUB AQUATIC TRAINING PTY LTD: David Melville, 2B Edgebaston Road, **Beverley Hills** NSW 2209

SUNDIVE CENTRE: Anthony Ross, East Esplanade, **Manly** NSW 2095

TERRIGAL DIVE SHOP: Les Graham, The Haven, **Terrigal** NSW 2260

The dive shop: Jim Prout, 656 Hunter Street, **Newcastle** NSW 2302

Tweed dive shop: Terry Morrison, 141 Wharf Street, **Tweed Heads** NSW 2485

United Divers: Leon & Dot Friend, 6 Victoria Street, **Wollongong** NSW 2500

Wilderness sea and ski: Jim Kovaks, 137 George Street, **Liverpool** NSW 2170

Wooden aquarium: Ross Wooden, 37 Morgan Street, **Wagga** NSW 2650

Victoria

Aquarius dive travel: Peter Stone, 40–42 Taylor Street, **Ashburton** VIC 3147

Associated divers: David Perry, 1288 Centre Road, **Clayton** VIC 3168

Ballarat skin ski and surf: John O'Neill, 89 Bridge Mall, **Ballarat** VIC 3350

Bob Cumberland's dive shop: Bob Cumberland, 34 East Concourse, **Beaumaris** VIC 3193

Dive and ski centre: Ray Walsh, 44 Beach Street, **Frankston** VIC 3199

Dive experience: John Marks, 82 Ferguson Street, **Williamstown** VIC 3016

Diveline scuba school: Gerry Scarfe, 214 Main Street, **Mornington** VIC 3931

Diversworld: Robbie Hiam, 23 Carrington Road, **Box Hill** VIC 3128

Diving headquarters: Ern Ireland, 436 High Street, **Prahran** VIC 3181

Diving instruction services: David Wailes, 12 Waratah Avenue, **The Basin** VIC 3154

Geelong divers: Graham Stevens, 146 Ormond Road, **East Geelong** VIC 3219

In-depth dive shop: Steve Sinclair, 580 Victoria Street, **North Melbourne** VIC 3051

John's dive shop: John Gibson, 300 St Kilda Road, **St Kilda** VIC 3175

Marlo dive and hire: Peter Hall, 59 Jorgensen Street, **Marlo** VIC 3888

Melbourne diving services: Dick Whittaker, 144 Bell Street, **Heidelberg** VIC 3081

Mornington dive centre: Jan & Brian Atkins, Shop 3, 10 Main Street, **Mornington** VIC 3931

North east divers: Gary Nash, 25A Rowan Street, **Wangaratta** VIC 3677

Ocean divers: Marty Owen, 237 East Boundary Road, **Bentleigh** VIC 3204

Portland aquanauts' fill station: Annette Hardy, PO Box 40, **Portland**, VIC 3305

Seaworld services: Anthony McMahon, Cnr Koroit & Banyan, **Warrnambool** VIC 3280

Ski and dive: The Manager, 1367 Nepean Highway, **Rosebud** VIC 3939

Southern cross divers: Russell Kitt, 1368 Toorak Road, **Burwood** VIC 3125

Victorian underwater services: Mike Slorach, 57 Waverley Road, **East Malvern** VIC 3145

Warrick's dive centre: Warrick Macdonald, 34 Station Street, **Moorabbin** VIC 3189

Warrick's dive centre: Warrick Macdonald, 167 Lonsdale Street, **Dandenong** VIC 3175

Western diving services: John Polwyk, 559 Mt Alexander Road, **Moonee Ponds** VIC 3039

Queensland

Adventure sports: Phil Feldman, 11 James Street, **Toowoomba** QLD 4350

Aren marine: Brooke Bayley, 222 Wishart Road, **Wishart** QLD 4122

Ausdive: The Manager, 5 Digger Street, **Cairns** QLD 4870

Barnes reef diving service: Greg & Charles Barnes, 153 Victoria Street, **Mackay** QLD 4740

Barrier reef diving services: Tony Fontes, PO Box 180, **Airlie Beach** QLD 4800

Bob Webb's dive shop: Bob Webb, 57 Cavendish Road, **Coorparoo** QLD 4151

Brisbane dive systems: Col McKenzie, 536 Rode Road, **Chermside** QLD 4000

Brisbane scuba school: Phil Browning, 2422 Logan Road, **8-Mile Plain** QLD 4123

Bundaberg dive centre: John English, 4B Targo Street, **Bundaberg** QLD 4670

Cairns barrier reef divers: Peter Boundy, 103 The Esplanade, **Cairns** QLD 4870

Cardwell dive: Shane Grant, C/- Cardwell Baths, **Cardwell** QLD 4816

Coral coast diving centre: Hayley Jorgensen, 55 Malcomson Street, **North Mackay** QLD 4740

Coral sea diving services: Brian Tweedie, PO Box 122, **Port Douglas** QLD 4871

Deep sea divers' den: Gill Hotchkiss, 319 Draper Street, **Cairns** QLD 4870

Diversion dive charters: Rainer Pabst, GPO Box 49, **Woody Point** QLD 4019

Diveventure scuba centre: David Saint, 3 Chappell Street, **Middlemount** QLD 4746

Diving academy: Chris Koch, 24 Goondoon Street, **Gladstone** QLD 4680

Downunder dive and sports: The Manager, Montipora Shops, **Airlie Beach** QLD 4802

Fish n fun: The Manager, Montipora Shops, **Airlie Beach** QLD 4802

H2O sportz: Tony Green, Hamilton Island, **Whitsunday Islands**, QLD 4700

Haven diving: Bob MacKinlay, PO Box 108, **Rockhampton** QLD 4700

Heron island dive shop: The Manager, Heron Island, **Via Gladstone** QLD 4680

Hervey bay dive shop: Glen Thompson, 429 Esplanade, **Torquay-Hervey Bay** QLD 4657

John Drury's scuba centre: John Drury, 36 Kirri Street, **The Gap, Brisbane** QLD 4061

Laver bros sports store: The Manager, 55 Goodoon Street, **Gladstone** QLD 4680

Mackay diving: Garry Roberts, 1 Mangrove Road, **Mackay** QLD 4740

Manta dive services: Bob King, 18 Yeldham Street, **Ingham** QLD 4850

Mareen world: George Boag, 8 Sydney Street, **Mackay** QLD 4740

MIKE BALL WATERSPORTS: Mike Ball, 252 Walker Street, **Townsville** QLD 4810

MITCHELL'S JETTY DIVERS' DEN: Wayne Boden, Loders Creek, Caravan Park, **Southport** QLD 4215

NOOSA SEA SPORT AND DIVE: Sid Vassallo, Hasting Street, **Noosa Heads** QLD 4567

NORTH COAST SPORTS & DIVE CENTRE: Bob Bilson, Shop 37 Kawana Water, **Mooloolaba** QLD 4557

PETER TIBBS PTY LTD: Peter Tibbs, PO Box 1983, **Cairns** QLD 4870

PRO-DIVE CAIRNS: The Manager, Marlin Jetty, Main Street, **Cairns** Qld 4870

REDLANDS SCUBA AND WATERSPORT: Barry Eldridge, 216 Old Cleveland Road, **Capalba** QLD 4157

ROBINSON'S SPORTS: Greg Riddel, 300 Queen Street, **Brisbane** QLD 4000

SCUBA WORLD BRISBANE: Ian McKinnon, 538 Milton Road, **Toowong** QLD 4066

SHUTE HARBOUR DIVING SERVICES: The Manager, Jubilee Pocket, **Shute Harbour** QLD 4802

SPORTSMAN'S WAREHOUSE: Sally Worrell, 32 Strathaird Road, **Bundall** QLD 4217

THE DIVE BELL: Col Hodgson, Shop 4–141 Ingham Road, West-End, **Townsville** QLD 4810

THE DIVE SITE: David Mann, Shop 3652, Ipswich Road, **Annerley** QLD 4103

TORQUAY SPORTS STORE: Glen Thompson, 429 Esplanade, **Torquay-Hervey Bay** QLD 4657

TOWNSVILLE SKINDIVERS: Fritz Quintemeyer, 95 Ingham Road, **Townsville** QLD 4810

WHALES SCUBA: Erroll & Nola Ansell, 44 River Esplanade, **Mooloolaba** QLD 4557

YEPPOON DIVE SHOP: Wayne Bailey, Shop 3/8 Normanby Street, **Yepoon** QLD 4703

South Australia

ADELAIDE SKINDIVING CENTRE: Paul Lunn, 7 Compton Street, **Adelaide** SA 5000

DIVERS INTERNATIONAL: Ian Stevens, 510 Goodwood Road, **Daw Park** SA 5041

DIVERS' SERVICE: Barrie Marsland, 80 Grange Road, **Welland** SA 5007

FLINDERS DIVING SERVICES: David Wilson, PO Box 1362, **Port Lincoln** SA 5606

NORTHERN UNDERWATER CENTRE: John & Heather Naumann, 158 Balmoral Road, **Port Pirie** SA 5540

PLOENGES MARINE & DIVING: Ian Ploenges, 343 Commercial Street, **Mt Gambier** SA 5290

PORT LINCOLN SKINDIVING CENTRE: The Manager, 73 Mortlock Terrace, **Port Lincoln** SA 5606

RAPID MARINE: Joe Csortan, 256 Rundle Street, **Adelaide SA** 5000

ROWE AND JARMAN SPORTS: The Manager, Shop 10 Westlands, **Whyalla** SA 5608

SEA HORSE SPORTS: The Manager, 1198 Grand Junction Road, **Hope Valley** SA 5090

Western Australia

AUSTRALASIAN DIVING CENTRE: Glen Parker, 259 Stirling Highway, **Claremont** WA 6010

BUSSELTON DIVERS: Ray Rodgers, 99 Queen Street, **Busselton** WA 6280

CRAIGIE SPORTS & TACKLE: The Manager, 11 Eddystone Avenue, **Craigie** WA 6025

DAFFY DUCK DIVING: Robyn Murphy, 7 Leon Road, **Dalkeith** WA 6009

DAVID BUDD WATERSPORTS: David Budd, 3 The Plaza, **Mandurah** WA 6210

DIVERS' EMPORIUM: The Manager, 33 Canning Highway, **East Fremantle** WA 6158

DIVERS' HIRE AND SUPPLY: Brian Lysaght, 71A Stirling Highway, **Fremantle** WA 6159

FREMANTLE SCUBA CENTRE: Michael Wong, 384 South Terrace, **South Fremantle** WA 6162

INDO PACIFIC DIVERS' WORLD: A Lavender, 143 Scarborough Beach Road, **Mt Hawthorn** WA 6016

INNERSPACE DIVING: Dave Culling, 281 Stewart Crescent, **Dampier** WA 6713

MALIBU DIVERS: Steve Sturgeon, 43 Rockingham Road, **Rockingham** WA 6168

MYAREE DIVING CENTRE: Edward Izydorski, Unit 5, 91 North Lake Road, **Myaree** WA 6154

MYREE SPORTS AND DIVING: Ian Riley, 6/93 Northlake Road, **Myree** WA 6156

PACE SPORTS: Trevor Beaver, 188 Marine Terrace, **Geraldton** WA 6530

PENGUIN DIVERS: Albert Collins, 46 Second Avenue, **Shoalwater** WA 6169

PERTH DIVING ACADEMY: Hugh Morrison, 281 Wanneroo Road, **Nollamara** WA 6061

PORT HEDLAND DIVE SUPPLIES: Vin Cooper, 3 Richardson Street, **Port Hedland** WA 6721

PRO-DIVE SERVICES: The Manager, 2 O'Brien Street, **Claremont** WA 6010

SCUBAQUATIC DIVING SCHOOL: Donna Summers, PO Box 42, **Dongara** WA 6525

SOUTH COAST DIVERS: Les Bail, 115 Stead Road, **Albany** WA 6330

SPORTS MARINE: The Manager, 117 Victory Street, **Bunbury** WA 6230

WA SKINDIVERS: Jack Sue, 486 Murray Street, **Perth** WA 6000

Tasmania

BICHENO DIVE CENTRE: Dave Warth, 4 Tasman Highway, **Bicheno** TAS 7215

KING MEADOW SPORTS: Kerry Luck, 120 Hobart Road, **Launceston** TAS 7250

LAUNCESTON DIVE CENTRE: Norm Bracken, 2 Trotters Lane, Prospect, **Launceston** TAS 7250

OCEAN SPORTS: Steve Pryde, 28 Waimea Avenue, **Sandy Bay** TAS 7005

SOUTHERN TASMANIAN DIVERS: Graham Blight, 212 Elizabeth Street, **Hobart** TAS 7000

TASMANIAN DIVING ACADEMY: Mike McGeary-Brown, 36 Corby Avenue, **West Hobart** TAS 7000

TASMANIAN SCUBA CENTRE: Bevin Brown, 62 Bass Highway, **Wynyard East** TAS 7325

TERRY CHARLTON SPORTS: Martin Daintith, 155–157 Brisbane Street, **Launceston** TAS 7350

THE DIVE SHOP: Chick Stanojevic, 233 Liverpool Street, **Hobart** TAS 7000

Northern Territory

TERRITORY DIVING SERVICES: Tim Proctor, Shop 9/2 Fannie Bay Place, **Fannie Bay** NT 5790

DIVING SITES AND HOW TO GET THERE

This appendix is restricted to sites mentioned in the text. For all diving information regarding sites and boats operating, the local professional dive shops listed in Appendix 1 can provide further assistance. Sites are listed according to the nearest town and follow the whole coastline from north Queensland round to Darwin.

TOWN	TYPE OF DIVING	SITES	SKILL	OPERATORS
Queensland				
Port Douglas	Charter and day trips	Outer Barrier Reef: Schnapper Is., Agincourt Reef, Undine Cay, Low Isles	Tourist, snorkle, scuba	Coral Sea Diving Services. **Boats:** *M V Quicksilver, Low Isles Cruises, Port Douglas*
Cairns	Charter and day trips	Coral Sea, Outer Barrier Reef Green Is., Michaelmas Cay, Hastings Reef, Lizard Is.	Tourist, snorkle, scuba	Green Island Resort Aquaflight, Air Queensland, Barrier Reef Divers, Deep Sea Divers Den. **Boats:** *Down Under, Hayles Cruises*
		Ribbon Reefs, Cod Hole, Euston Reef (by Charter only)	Scuba	**Charter boats:** *M V Reef Explorer, M V Auriga Bay, Bali Hai 11, Scheherazade, M V Divemaster*
Townsville	Charter and day trips	Magnetic Is., *Yongala* wreck, Mermidon Reef, *Foam* wreck	Tourist, snorkle, scuba	Mike Ball Watersports, Townsville. **Day boats:** *Reef Seeker, Coral Princess.* **Charter boat:** *Divemaster*
Mackay	Fly, dive and charter	Whitsunday Group, Hardy Lagoon	Snorkle, scuba	Air Whitsunday, Coral Coast Diving, Mackay Diving. **Boats:** *Elizabeth E Coral Cruises*
Bowen	Charter and day trips	Whitsunday Group	Snorkle, scuba	**Boat:** *Chiton 1*
Proserpine & Shute Harbour	Resorts, diving and charter	Hayman Is., Hook Is., Hamilton Is.	Scuba, snorkle	Barrier Reef Diving Services, Shute Harbour; Downunder Dive & Sports, Airlie Beach; H2O Sports, Hamilton Island. **Boats:** *Tafua, Solo, Seafari.*
	Bareboat charter	Whitsunday Group	Scuba, snorkle	Freedom ketches.
Rockhampton & Yeppoon	Charter and day trips	Keppel Is., Man & Wife Rocks	Snorkle, scuba	Haven Diving; Yeppoon Dive Shop. **Boats:** *Coralita, Worripa, Keppel Island catamarans*
Gladstone	Helicopter, day trip, charters	Heron Is., Capricorn Group, Lady Musgrave Is.	Tourist, snorkle, scuba	Heron Island Resort; Diving Academy; Laver Bros Sports. **Boats:** *Tropic Rover, Febrina, Triton 2, Safari, Vacationer*
Bundaberg & Town of 1770	Flights, day trips	Capricorn Group, Lady Elliot Is., Lady Musgrave Is.	Snorkle, scuba	Barrier Reef Airways; Bundaberg Dive Centre. **Boats:** *M V Lady Musgrave, Captain Cook.*
Brisbane	Day trips	Moreton Is., Tangalooma, Flinders Reef, Cowan Wreck Reef	Snorkle, scuba	Brisbane Dive Systems; Prodive; Diversion Dive Charters; John Drury's Scuba Centre; Tangalooma Resort
Gold Coast	Day trips	Stradbroke Is., Gold Coast, Cook Is., Mermaid Reef	Snorkle, scuba	Tweed Dive Shop; Terry Morrison

TOWN	TYPE OF DIVING	SITES	SKILL	OPERATORS
New South Wales				
Byron Bay	Day trips	Julian Rocks	Scuba	Byron Bay Dive Centre; Bill Sylvester Sports World
Coffs Harbour	Day trips	Solitary Is., North Solitary, South Solitary	Scuba	Dive Quest Divers, Mulloway; Solitary Island Divers Services; Coffs Harbour Dive Centre
South West Rocks	Day trips	Fish Rock Cave	Scuba	Mid Coast Divers; S W Rocks Marine Centre
Port Macquarie	Day trips	Cod Hole	Scuba	Mid Coast Divers, Port Macquarie
Lord Howe Island	Flights daily from Sydney	Lord Howe Island Lagoon, Balls Pyramid	Snorkle, scuba	Oxley Airlines, Lord Howe Island Resort
Forster	Day trips	Seal Rocks	Snorkle, scuba	Mid Coast Divers, Forster
Port Stephens	Day trips	Broughton Is./Halifax Park; Fly Point	Snorkle, scuba	The Dive Shop, Newcastle; Mid Coast Divers, Nelson Bay. **Boat:** *Calypso*
Sydney	Day trips	Harbour wrecks, offshore reefs, deep wrecks	Scuba	Dive 2000, Neutral Bay; Atlantis Divers, Palm Beach; Cronulla Dive Shop; (see Appendix 1 for others)
Jervis Bay	Day trips	Point Perpendicular, The Arch, docks, *Wandera* wreck, Smugglers' Cave, Bowen Is., Stony Creek	Scuba	Dive 2000, Sydney; Prodive, Huskisson; J B Seasport, Huskisson; Aquamedium, Canberra
Ulladulla	Day trips	Brush Is.	Scuba	Blue Marlin Divers
Narooma	Day trips	Montague Is.	Scuba	Darryl's Fishing and Dive Shop
Bermagui	Day trips	Montague Is.	Scuba	Montague Dive & Marine
Merimbula	Day trips	*Empire Gladstone* wreck	Scuba	Maurie Viewrow, Merimbula
Eden	Day trips	Green Cape wrecks	Scuba	Maurie Vierow, Merimbula
Victoria				
Wilson's Promontory	Day trips	Anser Is.	Scuba	Bob Cumberland's Dive Shop. **Boat:** Osprey
Melbourne	Day trips	Port Phillip, Portsea, Lonsdale Wall, Pope's Eye, Ship's Graveyard	Scuba	Southern Cross Divers; Sorrento; Diver Instruction Services. Blairgowrie; Western Diving Moonee Ponds; In Depth Dive (See Appendix 1 for others).
Moorabbin	Day trips	Beaumaris	Snorkle	Bob Cumberland's Dive Shop, Beaumaris
Geelong	Day trips	SS *Casino*	Scuba	Geelong Divers
Port Campbell	Local divers	*Loch Ard*	Scuba	Portland Aquanauts
Tasmania				
Hobart	Day trips	D'Entrecasteaux Channel, Bicheno Gulch	Scuba	Ocean Sports; Southern Tasmanian Divers; Dave Warth, Bicheno Dive Centre
Port Davey	Safari	S W Tasmania	Scuba	Remote area — no organised dives

TOWN	TYPE OF DIVING	SITES	SKILL	OPERATORS
South Australia				
Mt Gambier	Day trips	Piccaninnie Ponds,	Scuba, snorkle	Parks & Wildlife Permit, Mt Gambier
		Ewens Ponds	Snorkle, scuba	Department of Lands
		The Pines	Scuba	District forester, Tantanoola Reserve
		The Shaft	Scuba	Closed
		1080 Cave	Scuba	Manager, Barnoolut Station, PO Box 12, Mt Gambier; Ploenges Marine & Diving
(Permits from Cave Divers Association of Australia, PO Box 290, North Adelaide SA 5006, are required before landowners will give permission for scuba divers to enter.)				
Adelaide	Day trips	Port Noarlunga	Snorkle	Access from jetty
		Aldinga Reef	Scuba	Access from shore
		Victor Harbour	Scuba, snorkle	Adelaide Skin Diving Centre
Kangaroo Island	Day trips	Penneshaw jetty, *Loch Vennachar*	Scuba	Adventureland Diving Services, American Beach, Kangaroo Island
Western Australia				
Nullarbor Caves		Cocklebiddy Cave	Scuba	Remote area — no organised dives
Albany	Day trips	Whale chaser	Scuba	South Coast Divers, Albany
Busselton	Day trips	Cowaramup Bay	Snorkle	Shore Dive
		Canal Rocks	Scuba	Access from Cowaramup Bay
		Busselton jetty, Penguin Is., Five Fathom Bank wrecks	Scuba	Malibu Divers, Rockingham; Busselton Divers
		Alouran Is., Augusta	Scuba	
Perth	Day trips	Rottnest Is.	Tourists, scuba	Resort ferries from Perth
		Wreck trail	Snorkle, scuba	Rottnest Boatshed, Thompson's Bay; Western Diving Co, Perth; Jack Sue, Perth (See Appendix 1 for others).
		Cervantes	Snorkle	Shore diving
Bunbury	Shore dive	Lake Leschenaultia	Snorkle, scuba	Shire Ranger, Chidlow; Sports Marine, Bunbury
Geraldton	Charter	Abrolhos Is., *Batavia wreck*	Snorkle, scuba	Western Diving Co, Perth; Pace Sports, Geraldton; Subaquatic Diving School, Dongara. **Boats:** *Nellie Melba, Horizon*
Shark Bay	Shore dive	South Passage, Monkey Mia	Snorkle, tourist	Denham
Exmouth	Shore reef	Ningaloo	Snorkle	Innerspace Diving, Dampier
Broome	Charter	Ashmore & Cartier Islands, Rowley Shoals	Scuba, snorkle	Broome
Northern Territory				
Darwin	Safari	Coburg Peninsula	Snorkle, scuba	Territory Diving Services, Darwin

SAFETY PRECAUTIONS

1. Prospective scuba divers should ensure that their instructor is a professional. In Australia he/she will be a member of either PADI, FAUI or NAUI.
2. Divers should increase their skills by undertaking advanced and specialty diving courses.
3. Never dive alone. Ensure that your companion diver is adequately trained.
4. Ensure that your equipment is in good condition and that you are fully equipped with the necessary safety equipment.
5. Dive within the limits of your training and experience.
6. Ensure that diver charter vessels are 'in-survey' and that the vessel is equipped for safe diving practice with safety equipment including decompression bars and oxygen equipment.
7. Dive boat operators should be certified Dive Masters with rescue and first aid training.
8. Monitor depth, bottom time and available air supply during the dive.
9. Plan all dives within the NO DECOMPRESSION TABLES to allow the maximum safety factors. Add 3 m to depths indicated on gauge to allow for common gauge errors. Use the schedule for the next deeper depth and next longer time when calculating all dives.
10. Apply the safety RULE OF THIRDS — one third of your air for penetration or exploration, one third for return and one third in reserve for emergencies.
11. Ascend at designated rate of 18 m per minute. If you ascend at slower rates, bottom times should be reduced accordingly.
12. Sports divers should not exceed 39 m in depth, with 30 m being the recommended limit and 18 m the limit for novices.
13. Never dive when feeling unfit or ill. After an extended absence from diving, practise basic skills in calm conditions or book a refresher course with a dive operator.

FIRST AID AND EMERGENCY PROCEDURES

INJURY	EFFECTS	TREATMENTS
Shark or crocodile attack	Massive blood loss. Shock. Wound infection	Resuscitation on site after removing victim from water. Do not move victim to hospital until examined by doctor. No food. First aid brought to victim. Broad tourniquet or pressure bandage, or finger pressure on arteries. Pain killers, e.g. morphine and intravenous blood or plasma transfusion.
Striped catfish bite	Large specimens have caused death overseas. Pain may last for few hours to several days. Swelling, primary shock.	Application of pressure bandage or tourniquet. Immersion in hot water. Administration of emetine by doctor or congo red, methyl violet, dilute ascetic acid or potassium permanganate.
Stonefish bite	Death has occurred in Australia. Pain and oedema. Instantaneous shock.	As above. Antivenene available 2 ml ampoule — neutralises up to 2 stings.
Butterfly cod bite	Immediate intense pulsating pain, radiating in 3–10 minutes, subsides 3–8 hours. Nausea. Primary shock. Enlarged lymph glands.	As for striped catfish.
Box jellyfish sting	Death may occur within minutes. Pain in mounting waves. Purple or brown whip-like marks. Oedema, erythema and vesication lasting for 10 days.	Wear protective clothing. Application of tourniquet or pressure bandage. Douse with stingose (aluminium sulphate solution) or dilute acetic acid. External cardiac massage. Antivenene available.
Cone shell sting	Numbness. Respiratory paralysis.	Ligature between body and limb. Local incision of wound and venom removal. Artificial respiration.
Blue ringed octopus bite	Death has been known to occur. Paralysis within minutes. Breathing may cease.	Artificial respiration for 6–12 hours. Application of broad tourniquet and local incision over the bite.
Sea snake bite	Stiffness of limbs. Muscular paralysis. Breathing ceases.	Pressure bandage on limb. Local incision. Mouth to mouth resuscitation. Antivenene available.
Stingray sting	Severe laceration. Local pain.	Treat wound to prevent infection.
Decompression sickness	Pain near joints. Head-aches, double vision, paralysis, chest pains and nausea.	Administer 100% oxygen. Mouth to mouth resuscitation. Heart massage. Recompression chamber treatment.
Air embolism	Can cause severe brain damage or death. Forceful cry on exhalation when surfacing. Dizziness, convulsion, unconsciousness.	Immediate recompression in hyperbaric chamber.
Contaminated air	Carbon monoxide causes headaches, shortness of breath, vomiting and death.	Give 100% oxygen, mouth to mouth resuscitation, cardiac massage, hyperbaric oxygen if available.

INDEX

PHOTOGRAPHIC CREDITS

Front cover: K Deacon, Back cover: K Deacon, Back flap (top): Stephen Prigge, (bottom) Kath Savage, **1** A Fraser, **2-3** K Deacon, **4** Mike Gorman, **6** A Fraser, **8** Mike Gorman, **11** A Fraser, **12** A Fraser, **14** A Fraser, **16** K Deacon, **17a** K Deacon, **17b** K Deacon, **18** A Fraser, **19** Deacon Publishing Services, **20** K Deacon, **21** K Deacon, **22** K Deacon, **24a** J Frost, **24b** P Stone, **25** K Deacon, **26** K Deacon, **27a** K Deacon, **27b** K Deacon, **29a** A Fraser, **29b** K Deacon, **30** A Fraser, **31** K Deacon, **32** K Deacon, **34-35** K Deacon, **36a** K Deacon, **36b** K Deacon, **37a** K Deacon, **37b** BR Maloney, **38** Deacon Publishing Services, **39** K Deacon, **40** K Deacon, **42** A Fraser, **43** A Fraser, **44** K Deacon, **46** K Deacon, **47** K Deacon, **48** K Deacon, **49** C Deacon, **50** K Deacon, **51** K Deacon, **52-53** C Deacon, **54** K Deacon, **55a** K Deacon, **55b** B Pritchard, **56** K Deacon, **57** Deacon Publishing Services, **58** K Deacon, **59** C Deacon, **60a** A Fraser, **60b** K Deacon, **61** K Deacon, **62-63** A Fraser, **64a** K Deacon, **64b** K Deacon, **64c** K Deacon, **65** A Fraser, **66** A Fraser, **67** J Mullins, **68-69** C Deacon, **71** J Mullins, **72** K Deacon, **73** B Pritchard, **74** K Deacon, **75** A Fraser, **76** K Deacon, **78** K Deacon, **79** K Deacon, **80** A Fraser, **81** K Deacon, **82a** H Scarlett, **82b** K Deacon, **83** A Fraser, **85** A Fraser, **86a** H Scarlett, **86b** A Fraser, **87** D Warth, **88** B & P Boyle, **89** B & P Boyle, **91** C Deacon, **92** B & P Boyle, **93** A Fraser, **94** B & P Boyle, **95** K Deacon, **96** Deacon Publishing Services, **97** B & P Boyle, **98a** B & P Boyle, **98b** A Fraser, **99** B & P Boyle, **100** B & P Boyle, **101** B & P Boyle, **103** B & P Boyle, **104** D Warth, **105** B & P Boyle, **106** J Mullins, **107** K Deacon, **108-109** J Mullins, **110** J Mullins, **111** K Deacon, **112** K Deacon, **114** H Scarlett, **115** M Spencer, **117** K Deacon, **118** A Fraser, **119** K Deacon, **120** K Deacon, **121** K Deacon, **122** K Deacon, **123a** K Deacon, **123b** K Deacon, **125** A Fraser, **126** A Fraser, **127a** K Deacon, **127b** K Deacon, **127c** A Fraser, **128** K Deacon, **129** K Deacon, **130** K Deacon, **131** A Fraser, **132** K Deacon, **133a** Deacon Publishing Services, **133b** A Fraser, **136a** A Jolliffe, **136b** A Jolliffe, **137** B Heard, **138** A Jolliffe, **139** N Poulter, **140** N Poulter, **141a** A Jolliffe, **141b** A Jolliffe, **142** A Fraser, **144** BR Maloney, **145** Mitchell Library, **147a** K Deacon, **147b** D Lee, **148** K Deacon, **149** A Fraser, **150-151** K Deacon, **152** B & P Boyle, **153a** Deacon Publishing Services, **153b** Deacon Publishing Services, **154** K Deacon, **155** K Deacon, **156-157** K Deacon, **158** A Fraser, **159a** K Deacon, **159b** A Fraser, **160** K Deacon, **161** K Deacon, **162a** K Deacon, **162b** B & P Boyle, **163a** B Pritchard, **163b** K Deacon, **163c** A Fraser, **164** A Fraser, **165** B & P Boyle, **166** K Deacon, **167** I Oak, **168** A Fraser, **169a** J Dyer, **169b** K Deacon, **170-171** C Deacon, **173a** A Fraser, **173b** A Fraser, **174a** A Fraser, **174b** A Fraser, **175** K Deacon, **176-177** K Deacon, **178** K Deacon, **179** I Sizeland, **180** K Deacon, **181** A Fraser, **182** K Deacon, **183** A Fraser, **184** K Deacon, **186** Deacon Publishing Services, **187** Deacon Publishing Services, **188** A Fraser, **189** B & P Boyle, **190a** K Deacon, **190b** K Deacon, **191** K Deacon, **193** Deacon Publishing Services, **194** K Deacon, **195** Deacon Publishing Services, **196** M Spencer, **197a** M Spencer, **197b** M Spencer, **198-199** M Spencer, **200a** Deacon Publishing Services, **200b** Deacon Publishing Services, **200c** Deacon Publishing Services, **201** Deacon Publishing Services, **203** D Warth, **204** K Deacon, **205a** Deacon Publishing Services, **205b** Deacon Publishing Services, **206** K Deacon, **207** K Deacon, **209** M Spencer, **211a** Mitchell Library, **211b** Mitchell Library, **212** K Deacon, **213** B & P Boyle, **214** K Deacon, **215** A Fraser, **216-217** D Warth, **218** M Spencer, **219** K Deacon, **220** A Fraser, **222** K Deacon, **223a** A Fraser, **223b** A Fraser, **224** K Deacon, **225** A Fraser, **226** K Deacon, **227** A Fraser, **228-229** J Mullins, **230** A Fraser, **231** K Deacon, **232** A Fraser, **233** A Fraser, **234** K Deacon, **235** K Deacon, **236** A Fraser, **237a** K Deacon, **237b** B & P Boyle, **238-239** C Deacon, **241** K Deacon, **242** K Deacon, **243** A Fraser, **244** K Deacon, **245** A Fraser, **246** H Scarlett, **247a** Deacon Publishing Services, **247b** H Scarlett, **248** K Deacon, **249** K Deacon, **250-251** K Deacon, **252a** Deacon Publishing Services, **252b** Deacon Publishing Services, **252c** Deacon Publishing Services, **253** K Deacon, **254a** P Gibbs, **254b** Deacon Publishing Services, **255** K Deacon, **256-257** K Deacon, **258** K Deacon, **259** K Deacon, **260** A Fraser, **261** K Deacon, **262-263** A Fraser.